Plumbing

Cold Water Supplies, Drainage and Sanitation

F. Hall C.Ed, H.N.C. C.G.F.T.C.

Plumbing

Cold Water Supplies, Drainage and Sanitation

Longman
Scientific &
Technical

Longman Scientific & Technical
Longman Group UK Limited
Longman House, Burnt Mill, Harlow
Essex CM20 2JE, England
and Associated Companies throughout the world

First published 1981
Second edition 1986
Reprinted 1987

British Library Cataloguing in Publication Data
Hall, F.
    Plumbing: Cold water supplies, — 2nd ed.
    1. Plumbing   2. Buildings — Great
    Britain   3. Drainage, house — Great Britain
    I. Title
    696'.12'0941   TH6122

ISBN 0-582-98879-9

Produced by Longman Group (FE) Limited
Printed in Hong Kong

to Nigel

Preface

The installation of water supply, fire control, sanitary appliances and drainage systems forms an essential part of a building's services. It is therefore important that the installation of these systems is carried out economically and that they function efficiently when in use.

The craftsman plumber requires a knowledge of the principles on which these systems are designed and has to install them so that they satisfy the various Regulations and Codes of Practice. In order to ensure that the future craftsman plumber receives adequate training in modern techniques, the City and Guilds of London Institute and the various Regional Examining Bodies have prepared a syllabus for use in technical colleges throughout the country.

The primary aim of this book, therefore, is to assist the student plumber in his studies for the Advanced Craft Certificate and the text covers closely the requirements of the syllabus in advanced technology. Although the book is written specially for the plumbing student, it should also be useful to builders, surveyors and architects attending courses for the Technician Education Council's Diploma or Certificate. It should also be useful to students taking the Institute of Building examination in Building Services and Equipment or the City and Guilds examination in Building Services. Craftsmen plumbers and other personnel in the Building Industry should also find the book useful for reference purposes.

The plumbing student should also study hot water supplies, gas and oil installations, and the installation and maintenance of domestic heating systems and appliances. These subjects have been covered in a companion volume, *Plumbing: Hot Water Supply and Heating Systems—advanced craft*, by the same author, also published by Van Nostrand Reinhold.

In preparing this book, constant reference has been made to relevant Byelaws, Regulations and Codes of Practice. There may be regional differences in the Water Byelaws and the local water authority should therefore be consulted before designing and installing a water supply system in any particular area. Wherever possible, the written descriptions have been amplified by illustrations, tables and calculations.

I should like to thank the Publishers, Mr. C.R. Bassett for his encouragement and helpful criticisms, and Mrs. W. M. Whitney, librarian, for her help in obtaining various references from the Guildford County College of Technology, Department of Building and Surveying library. I should like to give very special thanks to my wife for her patience and understanding during the preparation of the book and for typing the entire manuscript.

F. Hall

Contents

Chapter 1
Water Supply

Rain Cycle

Water supply originates in nature in the form of rain, snow and hail falling from the clouds. Radiant heat from the sun causes evaporation of water on the earth's surface and the sea, thus forming clouds. The amount of water vapour that can be held by clouds depends on the temperature: when the temperature falls below the saturation point of the vapour, the clouds release the excess moisture, which falls to the earth. This process of evaporation and condensation is repeated and is known as the rain cycle.

Solvent Power of Water

Water has an extensive solvent capacity and there are very few substances that do not dissolve to some extent in water. Rainwater at the moment of formation is pure, but it soon ceases to be so. As rain falls through the atmosphere it dissolves some of the gases present, chiefly oxygen and carbon dioxide. It also washes the air so that the first fall is usually the most impure. By the time the water reaches the earth it contains various impurities and as it flows over or percolates through the soil it will dissolve other impurities present.

Sources of Water (Fig. 1.1)

Rainwater is soft and slightly acidic from the absorption of carbon dioxide (CO_2). On falling on moorland containing vegetable matter it dissolves more CO_2 and becomes distinctly, if weakly, acidic. This water will dissolve lead and is therefore known as plumbo-solvent. As the effects of lead poisoning are cumulative and very dangerous to health, such water must never be conveyed by lead pipes.

Shallow Wells. Shallow well water is obtained from sinkings in the top water-bearing strata of the earth. It should be treated with grave suspicion as it may become polluted from leaky drains or cesspools.

Intermittent or Land Springs. Since the water here is obtained from the same source as shallow well water it should be treated with the same grave suspicion.

Deep Wells. These are sinkings below the first impervious strata. Providing the well or borehole prevents the ingress of subsoil water, the water can usually be considered wholesome.

If the water passes through strata containing carbonate of calcium or magnesium, a certain amount of these salts will be dissolved in the water, depending on the amount of carbon dioxide present in the water. The water is now considered to be temporarily hard: this hardness can be removed by boiling the water, and can cause scaling of hot water pipes and boilers.

If the water passes through strata containing calcium sulphate, calcium chloride or

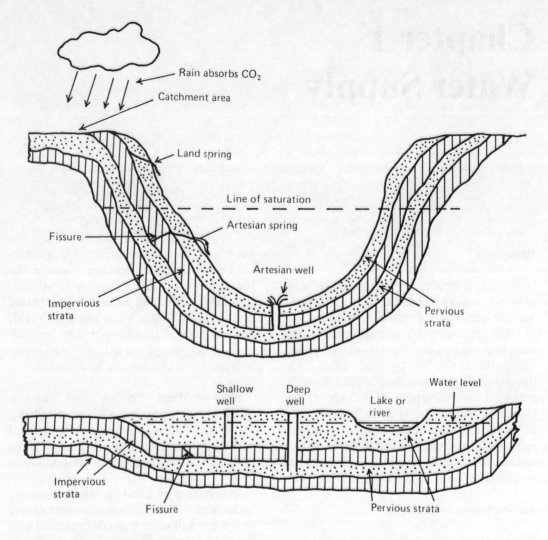

Rain absorbs CO_2

Catchment area

Land spring

Line of saturation

Artesian spring

Fissure

Artesian well

Impervious strata

Pervious strata

Shallow well

Deep well

Lake or river

Water level

Impervious strata

Fissure

Pervious strata

Fig. 1.1 Sources of water supply.

magnesium chloride, a certain amount of these salts will be dissolved in the water without the presence of carbon dioxide. This type of hardness cannot be removed by boiling the water and is termed permanent hardness. It will not cause scaling of hot water pipes and boilers, but it may cause corrosion.

Most waters contain both temporary and permanent hardness. The generally accepted classification of hardness is given in Table 1.1.

Artesian Wells and Springs. The water is obtained from the same source as deep well water and can be considered wholesome.

Table 1.1

Type of water	Hardness in parts per million
Soft	0 — 50
Moderately soft	50 — 100
Slightly hard	100 — 150
Moderately hard	150 — 200
Hard	200 — 300
Very hard	over 300

Lakes and Rivers. These are formed chiefly from the catchment of surface and

Table 1.2

Wholesome	1	Spring water	Very palatable
	2	Deep well water	
	3	Upland surface water	Moderately palatable
Suspicious	4	Stored rainwater	
	5	Surface water from cultivated lands	Palatable
Dangerous	6	River water to which sewage gains access	
	7	Shallow well water	

subsoil water. Water impounded in lakes from upland surfaces is soft and usually wholesome. River water is soft and generally turbid, especially after a storm: it may be rendered dangerous by discharges from factories and sewage works.

Classification of Water from Various Sources

The River Pollution Commissioners classify water from the various sources as shown in Table 1.2.

Analysis of Water

Two examinations should be made, namely for (a) bacteriological and (b) chemical content. The former is the more important from the health aspect, but the chemical content of the water must be considered because of its relation to attack on metals. If a new source of supply is to be used for

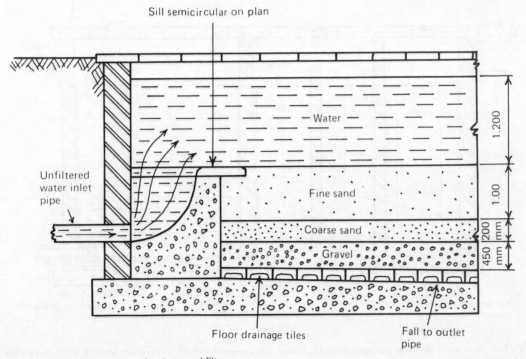

Sill semicircular on plan

Unfiltered water inlet pipe

Water

Fine sand

Coarse sand

Gravel

1.200

1.00

450 200 mm

Floor drainage tiles

Fall to outlet pipe

Fig. 1.2 Longitudinal section of a slow sand filter.

domestic purposes, or where doubt exists with an existing supply, the water should be submitted to an analyst for examination and report.

Filtration of Water

Slow Sand Filters. The simplest form of filter is one where the water passes downward by gravity through layers of sand and gravel (see Fig. 1.2). When the filter is first used it acts simply as a scourer, removing suspended matter but not removing harmful bacteria. In time, however, colloidal matter forms in the interstices between the sand grains. This gelatinous film is a barrier to the passage of harmful bacteria but gradually slows the passage of water to such an extent that it becomes necessary to remove the accumulated sludge. A certain amount of sand is removed in the process and it is therefore essential periodically to replace the surface with new sand. Fig. 1.3 shows a filter and storage cistern suitable for a small private water supply.

Pressure Filters. These consist of steel cylinders with the bottom filled with gravel and the remainder with sand (see Fig. 1.4). Water enters at the top and is collected in a perforated plate at the bottom, where there is a connection to an outlet pipe. The principle of operation is the same as the gravitational or slow sand filter, but since the water entering is under pressure the filtering process is much quicker. The efficiency of the filter is increased by adding a small dose of aluminium sulphate to the inlet water, which then forms a gelatinous film on top of the sand. The sand is cleaned by back washing and scouring with compressed air. The cylinder may be up to 2.7 m in diameter and the rate of filtration up to 12 m³ per m² of horizontal surface per hour.

Domestic Filters. The filter consists of an

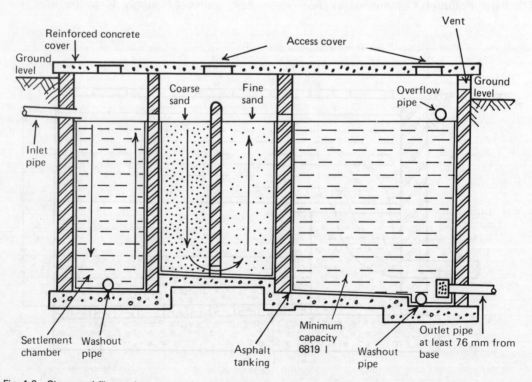

Fig. 1.3 Slow sand filter and storage cistern for a small private water supply. Note: water storage cistern maximum capacity 6820 l.

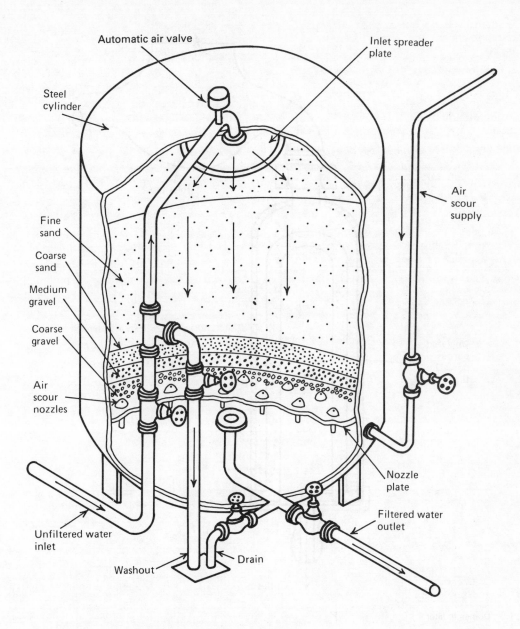

Automatic air valve

Inlet spreader plate

Steel cylinder

Air scour supply

Fine sand

Coarse sand

Medium gravel

Coarse gravel

Air scour nozzles

Nozzle plate

Filtered water outlet

Unfiltered water inlet

Washout

Drain

Fig. 1.4 Sectional view of Candy vertical-type pressure filter.

unglazed porous porcelain cylinder through which the water to be filtered flows. The filter is periodically cleaned in boiling water and impregnated with silver nitrate solution which has a sterilising effect upon the water. For passing larger volumes of water, filters may be obtained in batteries enclosed inside a cylinder. The type shown in Fig. 1.5 may be attached to any drinking water outlet tap.

Sterilisation of Water

In order to render large quantities of water safe for human consumption, sterilisation is required to destroy harmful bacteria.

Chlorine, because of its great efficiency when used in small quantities, is the most common reagent for the sterilisation of water. Its germicidal action in small doses is

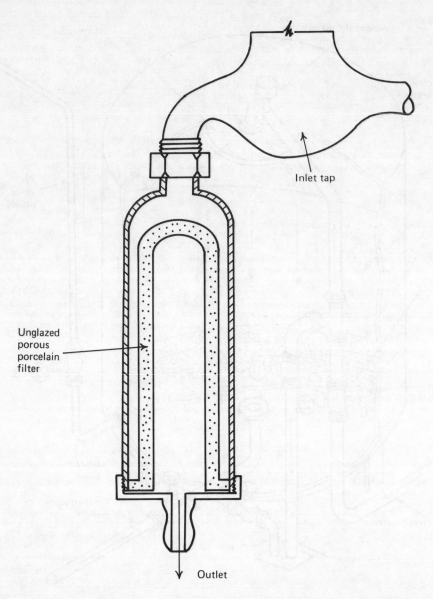

Inlet tap

Unglazed
porous
porcelain
filter

Outlet

Fig. 1.5 Domestic filter.

due to the destruction of enzymes necessary for the existence of microorganisms. It also has considerable oxidising powers, which favour the destruction of organic matter. The dosage of chlorine is strictly regulated so that there is sufficient to destroy any bacteria present but not enough to give an unpleasant taste to the water.

The chlorine is stored as a gas in steel cylinders from which it is injected into the water by automatic equipment. Fig. 1.6 shows details of a chlorinating plant which will automatically inject the correct amount of chlorine into a water main.

Water Softening

Hard waters are objectionable in domestic installations because more soap is required to produce a lather than when soft water is used. The term 'hardness' refers to the difficult of

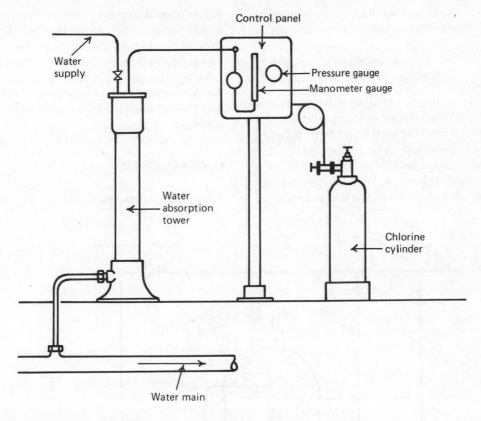

Fig. 1.6   Detail of chlorinating plant.

obtaining a lather with soap.

Permanently hard water may be softened by the use of sodium carbonate (washing soda), which causes the precipitation of calcium carbonate, leaving soluble sodium sulphate in solution.

Temporarily hard water may be softened by the use of slaked lime, which takes up the carbon dioxide from the bicarbonate present in the water, resulting in the precipitation of insoluble carbonate and the removal of temporary hardness. Slaked lime is used in conjuction with sodium carbonate in what is known as the *lime-soda process* of water softening.

Base-Exchange Process.   This process is used extensively for both industrial and domestic installations. The process removes both temporary and permanent hardness by passing the hard water through sodium zeolites. Sodium zeolites have the property of

exchanging their sodium base for the calcium or magnesium base, hence the term 'base exchange'. The process is as follows:

Sodium zeolite + Calcium carbonate
                              or sulphate
(inside   (in water)
softener)
becomes
Calcium zeolite + Sodium carbonate
                               or sulphate
(held inside   (in solution with
   softener)   the water)

After a period of use, the sodium zeolite is converted into magnesium and calcium zeolite, which has no softening power. It is therefore *regenerated* by the addition of common salt (sodium chloride). The salt is kept in contact with the magnesium and calcium zeolite for about half an hour, by which time the original sodium zeolite is produced. The process is as follows:

Calcium zeolite + Sodium chloride
(exhausted sodium)   (common salt)
becomes
Sodium zeolite + Calcium chloride
(regenerated)   (flushed to drain)

Fig. 1.7 shows a detail of a nonautomatic base-exchange water softener suitable for domestic use. The hard water to be softened passes downward through the zeolite. The value sequence for regeneration is as follows. *Backwash*: valves 1, 4, 5 closed; valves 2 and 3 open (also by-pass valve). *Salt addition*: all valves closed (except by-pass valve); salt added in specified dosage through the salt inlet at the top of the softener. *Salt rinse*: valves 2, 3 and 5 closed; valves 1 and 4 open (also by-pass valve). *After pre-set salt rinse period*: valve 4 and by-pass closed; valves 1 and 5 open, when the apparatus operates as a softener.

Other Impurities

Albuminoid ammonia indicates that organic

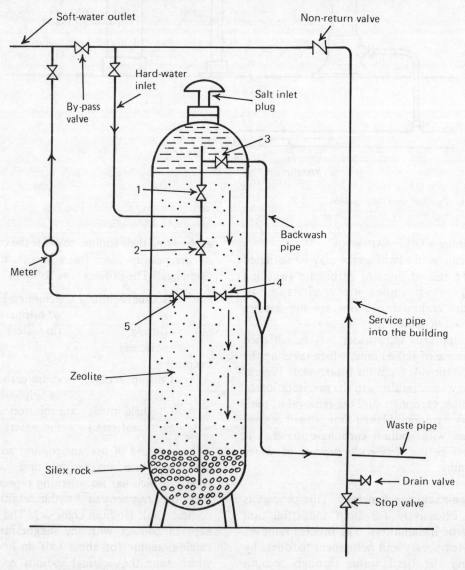

Fig. 1.7 Base-exchange water softener.

matter exists that is still undecomposed and gives the water a disagreeable taste and odour. Nitrites are a special danger as they indicate the presence of organic pollution. Nitrites represent the transitional stage in the oxidation of organic matter between ammonia and nitrates.

Nitrates signify past pollution and their presence without nitrites is an indication that the organic matter has been completely oxidised.

Water Quality

Water used for human consumption must be free from harmful bacteria and suspended matter, colourless, pleasant to taste, and for health reasons moderately hard.

Exercises

1. Sketch and describe the rain cycle.
2. Define
   (a) shallow well,
   (b) deep well,
   (c) artesian well,
   (d) spring.
3. Describe the characteristics of water from various supply sources.
4. Define
   (a) soft water,
   (b) temporary hardness of water,
   (c) permanent hardness of water.
5. State the two examinations required for water and explain the significance of each.
6. Sketch and describe
   (a) a slow sand filter,
   (b) a pressure filter,
   (c) a domestic filter.
7. Describe how water is sterilised.
8. Describe three methods of softening hard water.
9. Sketch a base-exchange water softener and describe its action.
10. What is the significance of the presence of albuminoid ammonia, nitrates and nitrates in water?
11. What are the characteristics of water for human consumption?
12. Define 'hardness' of water and describe how rainwater becomes 'hard'.
13. Sketch a section through an underground water storage tank.
14. State the advantages and disadvantages of slow sand filters and rapid filters.

Chapter 2
Cold Water Supply Systems

Definitions

The following definitions are used by most water authorities.

CISTERN: a container for water having a free water surface at atmospheric pressure.

FEED CISTERN: any storage cistern used for supplying cold water to a hot water apparatus.

STORAGE CISTERN: any cistern, other than a flushing cistern, having a free water surface under atmospheric pressure, but not including a drinking trough or drinking bowl for animals.

CAPACITY (of a cistern): the capacity up to the water line.

WATER LINE: a line marked inside a cistern to indicate the water level at which the ballvalve should be adjusted to shut off.

OVERFLOWING LEVEL: in relation to a warning or other overflow pipe of a cistern, the lowest level at which water can flow into that pipe from a cistern.

WARNING PIPE: an overflow pipe so fixed that its outlet end is in an exposed and conspicuous position and where the discharge of any water from the pipe may be readily seen and, where practicable, outside the building.

COMMUNICATION PIPE: any service pipe from the water main to the stop valve fitted on the pipe.

SERVICE PIPE: so much of any pipe for supplying water from a main to any premises as is subject to water pressure from that main, or would be so subject but for the closing of some stop valve.

DISTRIBUTING PIPE: any pipe for conveying water from a cistern, and under pressure from that cistern.

SUPPLY PIPE: so much of any service pipe which is not a communicating pipe.

MAIN: a pipe for the general conveyance of water as distinct from the conveyance to individual premises.

HOT WATER CYLINDER OR TANK: a closed container for hot water under more than atmospheric pressure. Note: a cylinder is deemed to include a tank.

POTABLE: water suitable for drinking.

FITTING: anything fitted or fixed in connection with the supply, measurement, control, distribution, utilisation or disposal of water.

Connection to Water Main

Before any building can be supplied with water from the main, it is essential to provide adequate notice in writing to the local water authority.

The tapping of the main and laying of the communication pipe is usually carried out by the local water authority at the building owner's expense. Where the local water authority permits a contractor to lay the communication pipe, the connection to the main will usually be made by the authority, also at the building owner's expense. Any underground piping should be inspected by the local water authority before being filled in.

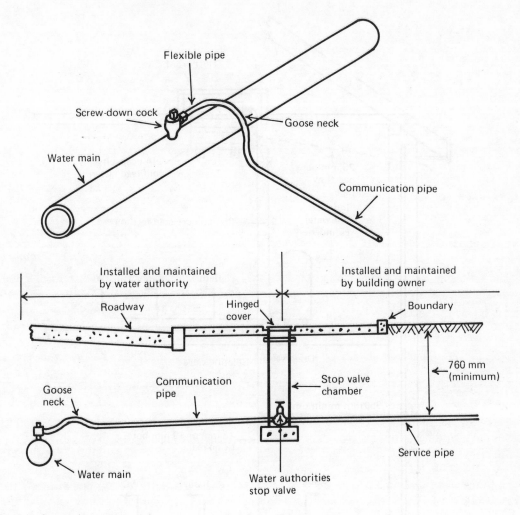

Fig. 2.1 Connection to water main.

In order to allow for any settlement of the communication pipe, a bend is made where the pipe connects to the main. Fig. 2.1 shows how the communication pipe is connected to the main and laid below ground.

Distribution Systems

Two distinct types of cold water system are used, depending on the water authority regulations.

Nonstorage or Direct Systems. In nonstorage or direct systems, all the sanitary fittings are supplied with cold water direct from the main. A cold water feed cistern is usually required to 'feed' the hot water supply system.

With certain types of electric or gas water heaters that may be supplied direct from the main, a cold water feed cistern is not required and this simplifies the system.

Fig. 2.2 shows a direct or nonstorage system for a house. The cold water feed cistern is small enough to be housed in the top of the airing cupboard, thus avoiding the risk of freezing.

Fig. 2.3 shows a direct or nonstorage system for a three-storey office or factory. The open outlet type gas or electric water heaters, which are connected direct from the main, avoid the use of a cold water feed cistern. A swivel outlet from the heaters may

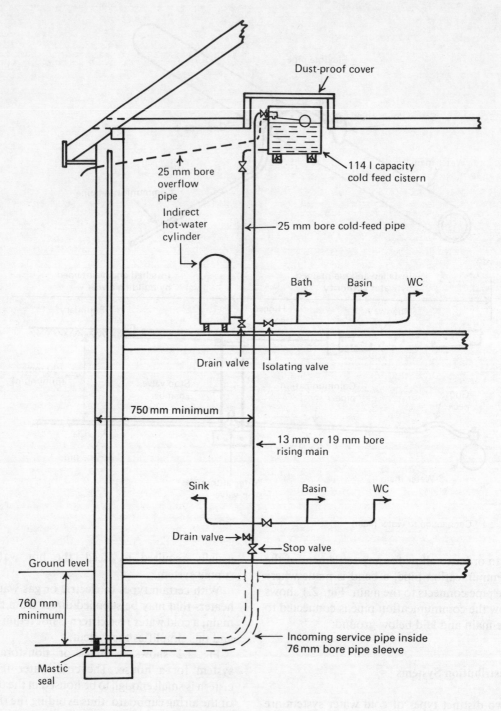

Dust-proof cover

25 mm bore overflow pipe

Indirect hot-water cylinder

114 l capacity cold feed cistern

25 mm bore cold-feed pipe

Bath   Basin   WC

Drain valve   Isolating valve

750 mm minimum

13 mm or 19 mm bore rising main

Sink   Basin   WC

Drain valve →   Stop valve

Ground level

760 mm minimum

Incoming service pipe inside 76 mm bore pipe sleeve

Mastic seal

Fig. 2.2   Direct system of cold water supply for a house.

be turned to supply hot water to each wash basin.

Indirect or Storage Systems. In indirect

or storage systems all the drinking water used in the building is supplied from the main and water used for all other purposes is supplied indirectly from a cold water storage cistern.

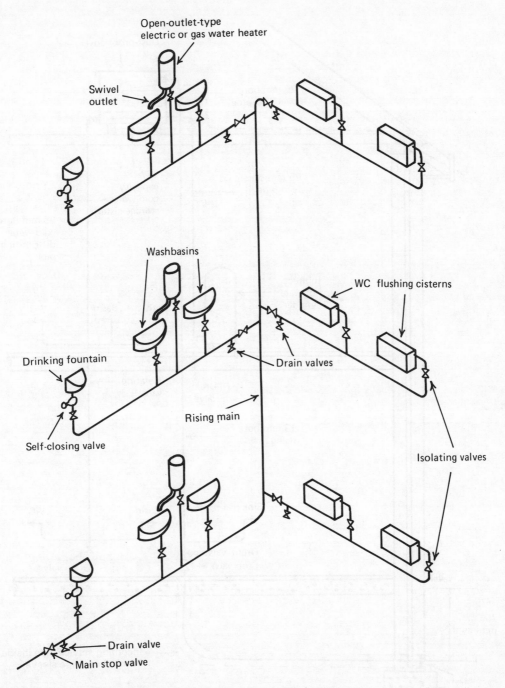

Fig. 2.3 Isometric diagram of a direct cold water system for an office or factory.

Since the cistern also supplies water to the hot water cylinder its capacity will be almost double the capacity required for the direct system.

Fig. 2.4 shows an indirect or storage system for a house. The cold water storage cistern is too large to be housed in the top of the airing cupboard and is therefore housed in the roof space near to the chimney breast to prevent freezing.

Fig. 2.5 shows an indirect or storage system for a three-storey office of factory.

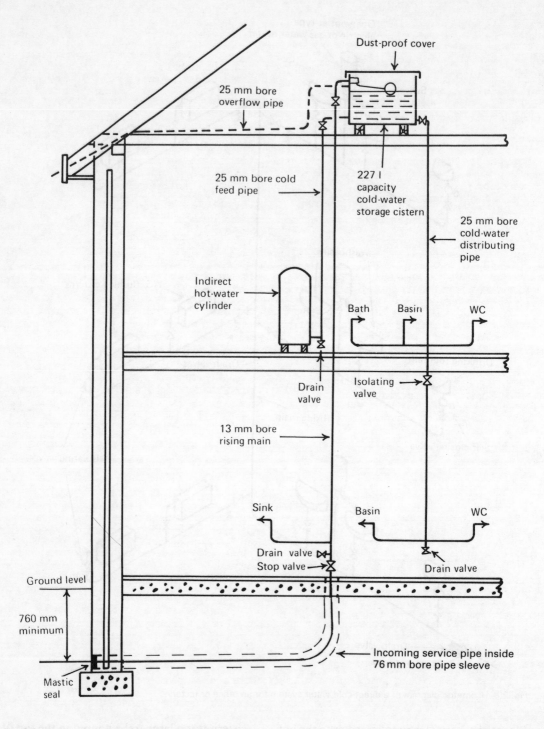

Dust-proof cover

25 mm bore
overflow pipe

25 mm bore cold
feed pipe

227 l
capacity
cold-water
storage cistern

25 mm bore
cold-water
distributing
pipe

Indirect
hot-water
cylinder

Bath   Basin   WC

Drain
valve

Isolating
valve

13 mm bore
rising main

Sink   Basin   WC

Drain valve
Stop valve

Drain valve

Ground level

760 mm
minimum

Mastic
seal

Incoming service pipe inside
76 mm bore pipe sleeve

Fig. 2.4   Indirect system of cold water supply for a house.

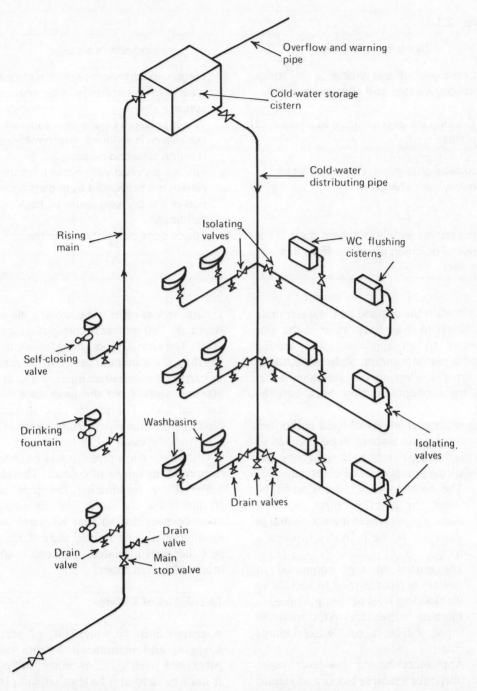

Fig. 2.5 Isometric diagram of an indirect cold water system for an office or factory

Advantages of direct and indirect cold water systems are given in Table 2.1.

Prevention of Back Siphonage

Back siphonage is the back flow of water, which may be contaminated, into the drinking water supply. In order for back siphonage to occur, a negative pressure or partial vacuum must be created in the pipe connected to an appliance having its outlet submerged in water, which may be contaminated. This is

Table 2.1

Direct or nonstorage	Indirect or storage
1. Less pipework and smaller or no cistern, making it easier and cheaper to install.	1. Large-capacity storage cistern provides a reserve of water during interruption of supply.
2. Drinking water is available at all draw-off points.	2. Water pressure on the taps supplied from the cistern is reduced, which minimises wear on taps and noise.
3. Smaller cistern, which may be sited below the ceiling.	3. Fittings supplied with water from the cistern are prevented from causing pollution of the drinking water by back siphonage.
4. In systems without a cistern there is no risk of polluting the water from this source.	4. Lower demand on the water main.

possible when the demand on the water main is sufficient to draw back water in the pipe connected to an appliance, thus leaving behind a partial vacuum. Siphonic action is then created allowing contaminated water from the appliance to flow back into the main.

The water regulations of local authorities require cold water systems to be installed so as to prevent the risk of back siphonage and the following points must be observed:

1. The ballvalves in cisterns must be above the overflow pipe and if a silencer pipe is fitted it must discharge water above the ballvalve through a spray.
2. The outlets of taps connected to sanitary appliances must be well above the flooding level of the appliance.
3. Flushing valves for WCs must be supplied from a cold water storage cistern.
4. Appliances having low-level water inlets, for example bidets and certain types of hospital appliance, must be supplied from a cold water storage cistern and never direct from the main.

Precautions Against Frost Damage

Water regulations generally require under-

ground service pipes to be laid at a minimum depth of 760 mm as a precaution against frost. The service pipes should also enter the building at a minimum depth of 760 mm and be carried through to an internal wall at least 600 mm away from the inner face of any external wall. The service pipe should run directly up to the cistern and be kept at least 2 m from the eaves.

Cistern overflow pipes should be arranged to prevent the inflow of cold air. This can be achieved by terminating the pipe about 50 mm below the water line. Drain valves must be provided so that all parts of the installation may be drained and cisterns must be either well insulated or placed inside an insulated cistern room.

Installation of Cisterns

A cistern must be watertight, of adequate strength, and manufactured from plastic, galvanised steel, asbestos cement or copper. It must be sited at a height that will provide sufficient head and discharge of water to the fittings supplied. It must be placed in a position where it can be readily inspected and cleansed. A cistern must be provided with a dust-proof but not airtight cover and protected from damage by frost.

Every storage cistern must be fitted with an efficient overflow pipe which should have a fall as great as practicable not less than 1 in 10. In England and Wales the regulations require the overflow pipe to have an internal diameter greater than the inlet pipe and in no case less than 19 mm. In Scotland the regulations require an overflow pipe to have an internal diameter of not less than twice that of the inlet pipe and in no case less than 32 mm.

If the capacity of a storage cistern does not exceed 4546 litre, the overflow pipe should be arranged as a warning pipe, i.e. so that its outlet is in a conspicuous position, either inside or outside the building, so that any discharge of water can be readily seen. There should be no other overflow pipe. If the capacity of a storage cistern exceeds 4546 litre it should have a warning pipe as previously described, or alternatively it should have an overflow pipe not arranged as a warning pipe and, in addition, a warning pipe of not less than 25 mm internal diameter, or some other device, which effectively indicates when the water reaches a level not less than 50 mm below the overflowing level of the overflow pipe.

Fig. 2.6 shows the method of installing a cistern in a roof space.

Duplication of cisterns It is sometimes difficult to decide when to duplicate cisterns. As a guide, duplication is usually necessary when the storage exceeds 4500 litre. Two or more cisterns interconnected so that each cistern may be cleansed or renewed without cutting off the supply of water from the remaining cisterns will be re-

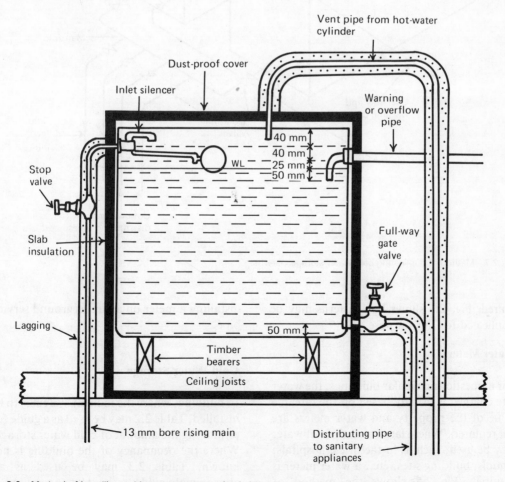

Fig. 2.6   Method of installing cold water storage or feed cistern.

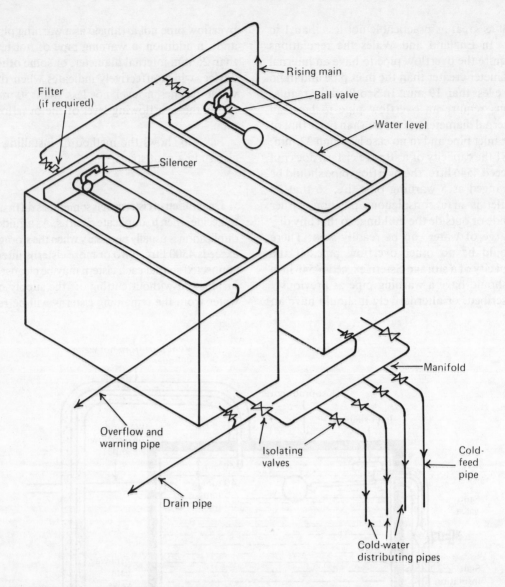

Fig. 2.7 Method of duplicating cold water storage cisterns.

quired. Fig. 2.7 shows how cisterns may be duplicated for this purpose.

Water Meters

For domestic and similar buildings, the water rate charged is usually based on the rateable value of the property and water meters are not required. Where large quantities of water may be used, such as in factories, hospitals, schools, building sites, etc., a water meter is required. Fig. 2.8 shows the method of

installing a meter on an underground service pipe.

Cold Water Storage

In buildings where the indirect system is to be installed, Table 2.2 may be used as a guide for estimating the amount of cold water storage. Where the occupancy of the building is not known, Table 2.3 may be used as an approximate guide to the storage required.

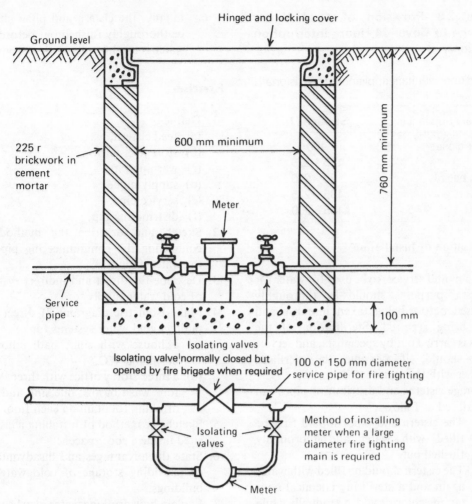

Fig. 2.8 Method of installing a meter on the underground service pipe.

Table 2.2 Provision of Cold Water Storage to Cover 24 Hours Interruption of Supply

Based upon occupancy, reproduced by permission of the British Standards Institution.

Type of building		Storage (1)	Type of building		Storage (1)
Dwelling houses and flats	per resident	90	Offices with canteens	per head	45
Hostels	per resident	90	Restaurants	per head/per meal	10
Hotels	per resident	140	Day schools	per head	30
Offices without canteens	per head	40	Boarding schools	per head	90
			Nurses' homes and medical quarters	per resident	115

19

Table 2.3 Provision of Cold Water Storage to Cover 24 Hours Interruption of Supply

Based upon sanitary appliances (provisional).

Sanitary appliance	Storage (1)
Water closet	180
Sink	135 – 225
Wash basin	90 – 250
Shower	135 – 225
Urinal	135 – 250

Sterilisation of Installation

All mains and services to be used for water for domestic purposes should be thoroughly sterilised before being taken into use and after being repaired. Sterilisation of the mains is carried out by specialists, and service pipes should, if possible, be sterilised together with the mains.

Storage cisterns and distributing pipes can be sterilised as follows:

1. The cistern and pipes should first be filled with water and thoroughly flushed out.
2. The cistern should be filled with water again and a sterilising chemical containing chlorine added gradually while the cistern is filling. Sufficient chemical should be added to give the water a dose of 50 parts of chlorine to one million parts of water.
3. When the cistern is full, the supply should be stopped, and all the taps on the distributing pipes opened successively, working progressively away from the cistern. Each tap should be closed when water discharged smells of chlorine.
4. The cistern should then be topped up with water from the service pipe pipe and more chlorine added.
5. The cistern and pipes should then remain charged for at least three hours and then tested for residual chlorine; if none is found, the sterilisation will have to be carried out again.
6. Finally, the cistern and pipes should be thoroughly flushed out before any water is used for domestic purposes.

Exercises

1. Define
   (a) feed cistern,
   (b) storage cistern,
   (c) warning pipe,
   (d) supply pipe,
   (e) service pipe,
   (f) distributing pipe.
2. Sketch and describe the method of connecting the communicating pipe to the water main.
3. Describe the direct and indirect systems of cold water supply.
4. Make isometric diagrams of direct and indirect cold water systems for
   (a) a house with sink, bath, shower, basin and WC,
   (b) a three-storey office with three WCs, four wash basins, one sink and one drinking fountain on each floor.
5. Sketch the method of installing a cistern sited inside a roof space.
6. State the advantages and disadvantages of providing storage of cold water in buildings.
7. Explain how drinking water may become polluted by back siphonage and the methods used for its prevention.
8. Sketch the method used to interconnect two or more cisterns to provide a means of repairing, renewing or cleaning one of the cisterns whilst still maintaining a supply to the building.
9. Describe the precautions necessary in installing a cold water system to prevent damage to the system by the action of frost.
10. Sketch the method of installing a meter and state the types of building requiring a meter.
11. State the method used to estimate the amount of cold water storage in buildings.
12. Describe how a cold water system is sterilised.

Chapter 3
Boosted Cold Water Installations

Principles

For building constructed so that the water supply to them is above the level of the mains head, it will be necessary to provide boosting equipment. This problem arises in multi-storey buildings or buildings constructed on high ground.

If, for example, the pressure on the water main during the peak demand period is found to be 400 kPa, this pressure would supply water inside the building up to a height of about 40 m. In order to provide a good delivery of water at the highest fittings a residual head of 2 m above these fittings is usually required. In this example, therefore, the main would supply water up to a height of $40 - 2 = 38$ m.

If the building is multi-storey with a vertical height from the main to the highest fittings of, say, 50 m, it will be necessary to provide boosting equipment to lift the water from 38 m to 50 m.

There are basically four systems in common use:

1. directly boosted system to cold water cisterns only,
2. directly boosted system to a header pipe,
3. indirect boosting from break cisterns at low level,
4. the auto-pneumatic system.

Directly Boosted System to Cisterns (Fig. 3.1)

Where the water regulations permit, pumps can be connected directly to the incoming main, thus enabling the pump head to be added to that of the mains. Control of the pump is effected by means of a float switch or electrode probes in the roof-level storage and drinking water header cistern. Fig. 3.2 shows a detail of the drinking water header cistern.

Directly Boosted System to Header Pipe (Fig. 3.3)

This system incorporates an enlarged header pipe to provide drinking water at high level between pump on-off cycles. The storage cisterns are provided with a float or electrode probes and the refilling of the header pipe is controlled by a pipeline switch (see Fig. 3.4). A time delay of approximately 2 min is fitted to the pipeline switch. As the header pipe is normally refilled in a shorter time than this, the excess water is discharged into the storage cisterns.

Centrifugal pumps, however, can operate against a closed valve for a short period of time and therefore no damage to the pumps is likely to occur if the ballvalves are closed during the refill cycle.

Indirect Boosting from Low-Level Cistern

Many water authorities require a break cistern to be installed between the main and the boosting unit. The cistern will serve as a

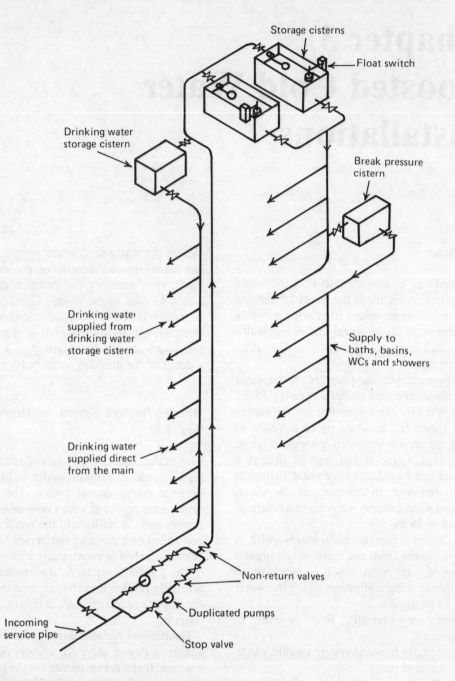

Storage cisterns

Float switch

Drinking water
storage cistern

Break pressure
cistern

Drinking water
supplied from
drinking water
storage cistern

Supply to
baths, basins,
WCs and showers

Drinking water
supplied direct
from the main

Non-return valves

Incoming
service pipe

Duplicated pumps

Stop valve

Fig. 3.1   Directly boosted to drinking water and storage cisterns.

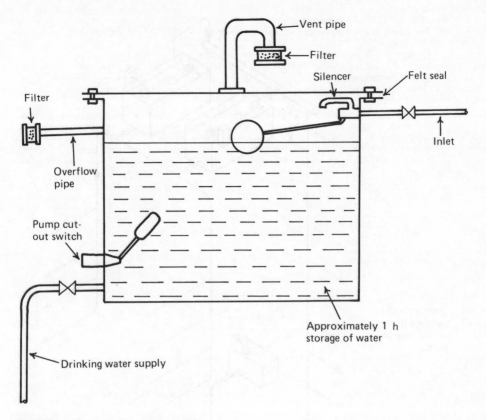

Fig. 3.2   Detail of drinking water header cistern.

boosting reservoir and prevent lowering of the pressure on the main. As there is no assistance from the mains pressure, the boosting equipment must be capable of overcoming the total static head of water plus the frictional resistances in the pipework.

The sizing of the break cistern must be considered carefully to prevent stagnation of water, which could occur because of oversizing. Where the whole of the storage is to be provided at low level, the local water authority must be consulted.

Float or electrode probe control switches must be provided to the break cistern to switch off the pumps when the water level drops to about 250 mm above the pump suction inlet. This precaution is necessary to prevent the pumps running dry during an interruption of supply. Fig. 3.5 shows details of the low-level break cistern.

Auto-pneumatic System (Fig. 3.6)

In an indirect system of cold water supply, a steel cylinder may be used as an alternative to a drinking water header cistern or pipe. The cylinder contains compressed air in the top which is pressured by the water pumped into the bottom. This cushion of air serves to force water up to the high-level drinking water points and storage cistern. When drinking water is drawn off through the high-level fittings, the water level in the cylinder falls. At a predetermined low level a pressure switch cuts in the pump and the cylinder is refilled up to a predetermined high level, when another pressure switch cuts out the pump.

Purpose of the Air Compressor.   In time some of the air inside the cylinder is absorbed into the water and a gauge glass is usually fitted to give a visual indication of the water level. As the air becomes absorbed, a smaller quantity is available to provide the required reserve of pressure and the frequency of

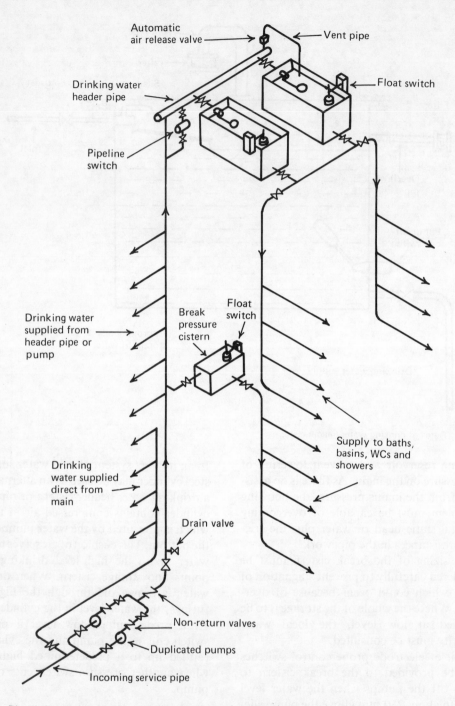

Fig. 3.3   Directly boosted to drinking water header pipe and storage cisterns.

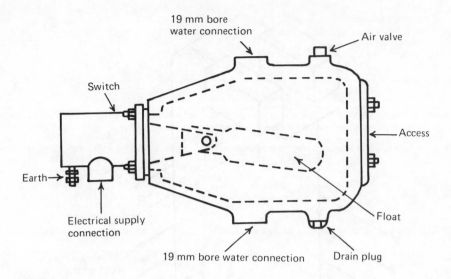

19 mm bore
water connection

Air valve

Switch

Access

Earth

Float

Electrical supply
connection

19 mm bore water connection

Drain plug

Fig. 3.4　Pipeline switch.

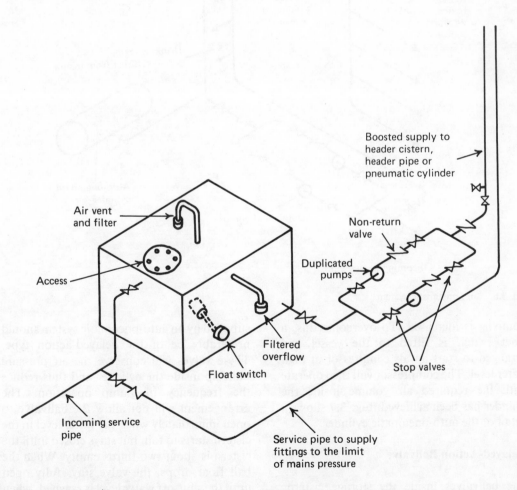

Boosted supply to
header cistern,
header pipe or
pneumatic cylinder

Air vent
and filter

Non-return
valve

Access

Duplicated
pumps

Filtered
overflow

Float switch

Stop valves

Incoming service
pipe

Service pipe to supply
fittings to the limit
of mains pressure

Fig. 3.5　Low-level break cistern.

25

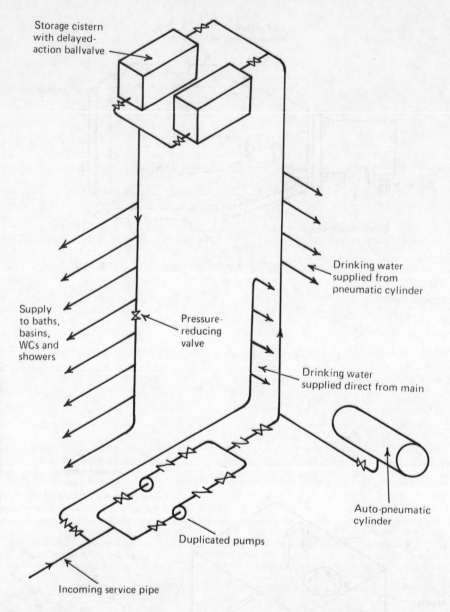

Storage cistern
with delayed-
action ballvalve

Drinking water
supplied from
pneumatic cylinder

Supply
to baths,
basins,
WCs and
showers

Pressure-
reducing
valve

Drinking water
supplied direct from main

Auto-pneumatic
cylinder

Duplicated pumps

Incoming service pipe

Fig. 3.6 Auto-pneumatic system.

pumping is increased. To overcome this, a float switch is fitted to the vessel and arranged to start an air compressor at high water level. The compressor will then operate until the required air volume inside the cylinder has been achieved. Fig. 3.7 shows a detail of the auto-pneumatic cylinder.

Delayed-Action Ballvalve

The ballvalves inside the storage cisterns

supplied by an auto-pneumatic system should preferably be of the delayed-action type. These valves will conserve the air pressure built up inside the cylinder and thus reduce the frequency of pump operation. The arrangement will not allow the ballvalve to open immediately when the water level in the cistern starts to fall, but stays closed until the cistern is about two-thirds empty. When the ball float drops, the valve stays fully open until the shut-off water level is reached, when

26

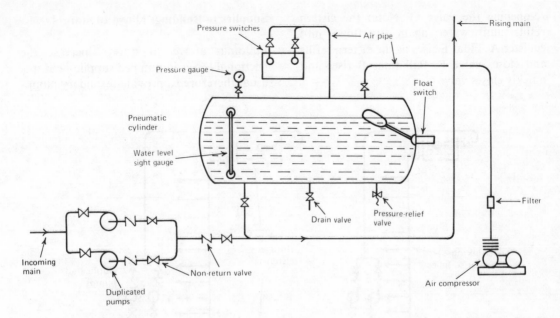

Fig. 3.7 Automatic pneumatic cylinder.

it closes quickly.

Fig. 3.8 shows the sequence of operation of the valve. Fig. 3.8a shows the cistern full, valve F closed and the water flowing over the rim into the ball float canister A. The ball float B is rising and closing the inlet valve C. In Fig. 3.8b, the cistern is being emptied and the ball float B is floating on the water, which remains in the canister A. Valves F and C remain closed. In Fig. 3.8c, water has been drawn off from the cistern to a predetermined level, which allows float E to be lowered. Valve F is now opened, which releases water from canister A and lowers ball float B thus

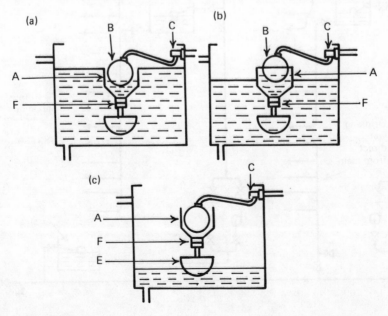

Fig. 3.8 Delayed-action ballvalve.

re-opening the valve C. Note: the cistern refills until water again overflows into canister A. Float E rises as the cistern refills and closes valve F. Ball float B rises and quickly closes valve C.

Supplies to Buildings above 20 Storeys

Buildings above 20 storeys increase the frictional losses in pumped supplies and it is usual therefore to provide secondary pump-

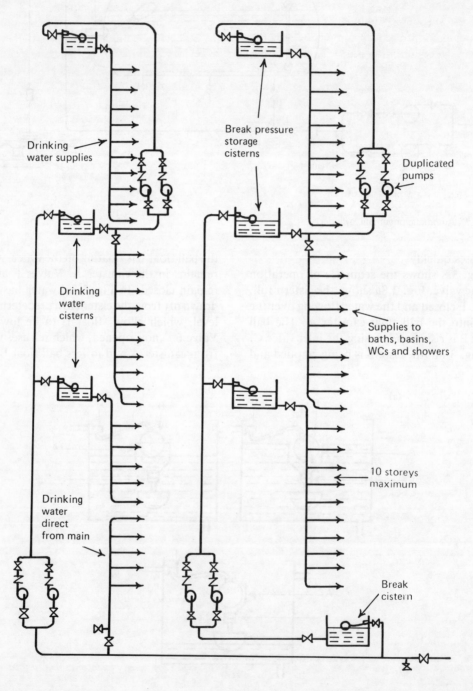

Fig. 3.9   System for 30 storeys.

ing equipment on the twentieth floor. Fig. 3.9 shows a system for 30 storeys.

Distribution from Storage Cisterns

In order to avoid excessive pressures in the pipework, the maximum head of water in the system must be limited to 30 m. The floors of a multi-storey building must therefore be zoned by means of a break-pressure cistern or pressure-reducing valve. Figs. 3.1, 3.3 and 3.9 show the method of zoning by break-pressure cisterns and Fig. 3.6 shows zoning by use of a pressure-reducing valve.

Exercises

1. Sketch sections through the following and describe their operation:
   (a) break cistern,
   (b) delayed action ballvalve,
   (c) pipeline switch.
2. Sketch and describe the operation of a cold water installation for a 16-storey building when the water is to be pumped:
   (a) directly from the main,
   (b) indirectly from the main.
3. Sketch a section through a pneumatic cylinder and describe its operation.
4. State the purpose of the following and explain their operation:
   (a) break cistern,
   (b) break-pressure cistern,
   (c) float switch,
   (d) automatic air valve.
5. Sketch and describe a cold water installation for a 30-storey building.
6. Sketch a method of interconnecting two pumps so that one of the pumps may be used for stand-by purposes.

Chapter 4
Fire Control Systems

Hose Reels

These may be used by the occupants of the building as a first aid and it is often possible to extinguish a fire by a jet of water before the fire brigade arrives. It is, however, possible to extinguish a fire by means of a suitable portable extinguisher and for this reason these should not be dispensed with when hose reels are installed.

Positioning. As hose reels are intended for use by the occupants of the building, they should be sited in positions that will be accessible without exposing the user to danger from the fire. For this reason they are usually fixed along escape routes or close to the fire exits, so that personnel escaping from a fire will pass them on their way to safety and may use them without having their means of escape cut off.

In office blocks, especially the multi-storey type, the hose reels must be fitted inside the actual office accommodation, which means that they are usually fitted adjacent to the fire exit doors. This enables the hose reels to be used without opening the smoke stop doors of the escape lobby and therefore prevents the lobby being filled with smoke.

In industrial buildings it is not always possible to fix hose reels only near to the exit doors, owing to the fact that the width of the building may prevent the hose nozzle reaching a fire in the centre. In these buildings it is also necessary to fix additional hose reels in the centre of the building, usually on the columns.

Design Considerations. Hose reel installations should be designed so that no part of the floor is more than 6 m from the hose nozzle when the hose is fully extended. The water supply must be able to provide a discharge of not less than 0.4 l/s through the nozzle and also designed to allow not less than three hose reels to be used simultaneously at a flow rate of 1.2 l/s. A water pressure of 200 kPa is required at the nozzle and with this pressure the jet will have a horizontal distance of 8 m and a height of about 5 m.

Pipe Size. The Fire Offices Committee Rules require that pipes supplying hose reels should be not less than 50 mm in diameter and that the connection to the hose reel should not be less than the nominal bore of the hose.

It is usual to use a 50 mm diameter pipe for buildings up to 15 m in height and a 64 mm diameter pipe for buildings above 15 m in height. In some areas, the minimum diameter of the pipe connected to each hose reel should not be less than 25 mm in diameter.

Water Supply. If the supply authority's main can supply a minimum pressure at the highest reel of 200 kPa and also provide sufficient discharge of water, the hose reels may be supplied directly from the main (see Fig. 4.1).

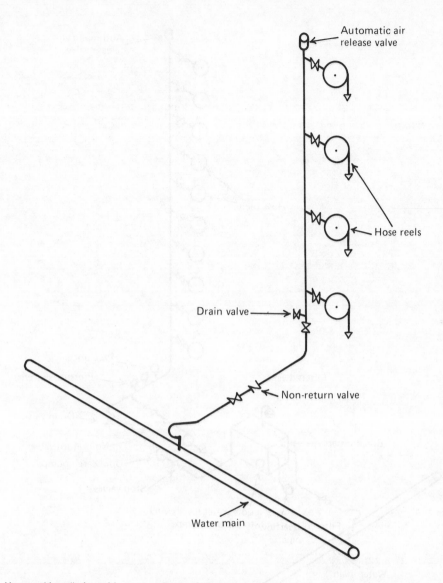

Fig. 4.1 Hose reel installation without pumping equipment.

If, however, the supply authority's main will not satisfy the required conditions, automatic pumping equipment will have to be installed. Some authorities will allow pumping equipment to be connected directly to the water main, provided that a reasonable flow would take place under mains pressure at the highest hose reel without pumping.

If the water supply authority requires a break cistern, this should hold a minimum of 1.6 m³ and duplicated pumps should be installed that will provide a minimum discharge of 2.3 l/s. In large buildings a stand-

by pump operated by a diesel engine may be required.

Fig. 4.2 shows a hose reel installation with pumping equipment. When a reel is used, the drop in water pressure allows one of the pressure switches to switch on the duty pump. As an alternative to a pressure switch, a flow switch inserted in the main on the delivery pipe from the pump may be used. The switch is capable of sensing a flow of 0.1 l/s and will keep the duty pump running until the hose reel is shut off.

Fig. 4.3 shows the installation of a fixed-

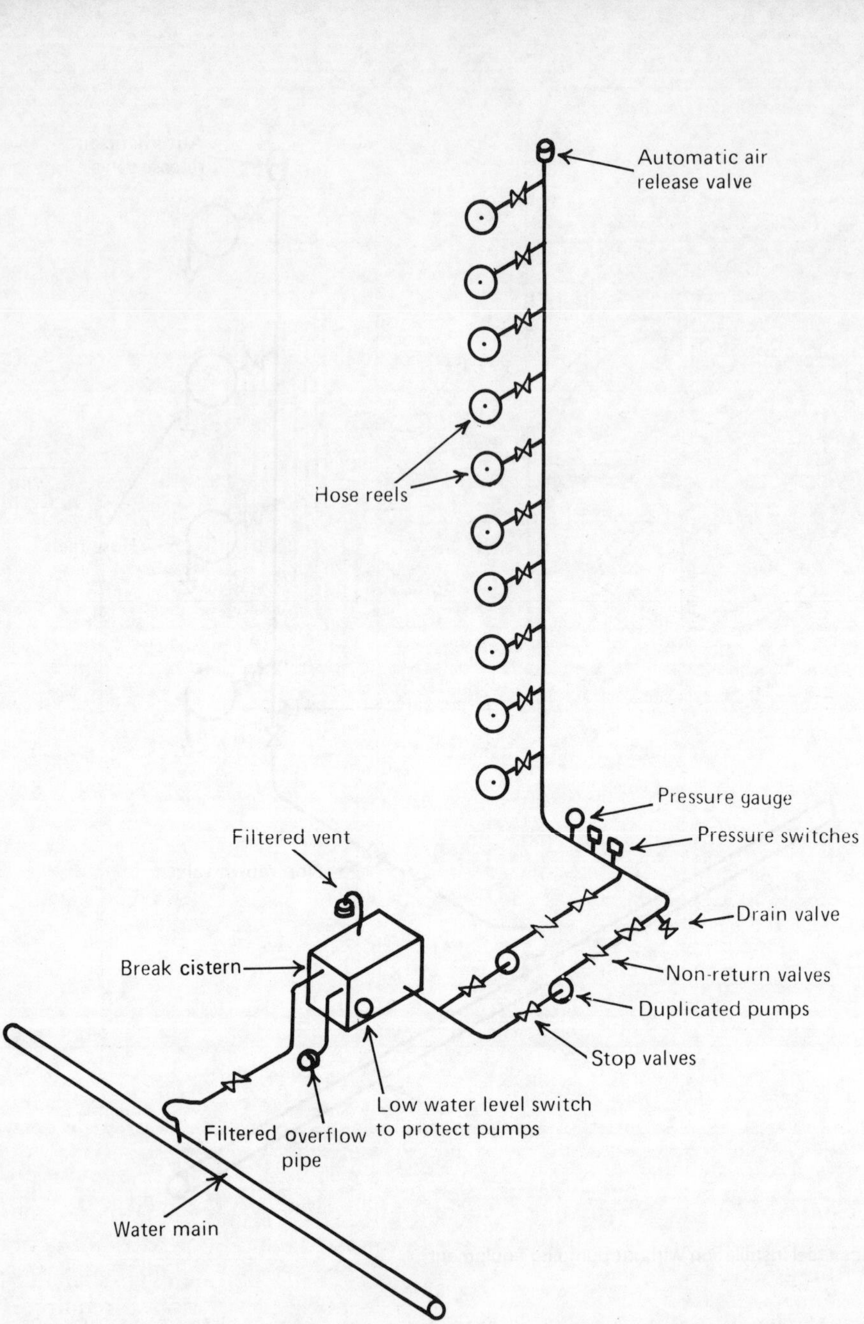
Automatic air release valve
Hose reels
Pressure gauge
Filtered vent
Pressure switches
Drain valve
Break cistern
Non-return valves
Duplicated pumps
Stop valves
Low water level switch to protect pumps
Filtered overflow pipe
Water main

Fig. 4.2 Hose reel installation with pumping equipment.

type hose reel: a swinging-type hose reel may also be installed inside a corridor recess.

Dry Riser

A dry riser consists of an empty pipe rising vertically inside a building with landing valves connected to it on each floor and at roof level. An inlet is provided at ground level to allow the fire brigade to pump water into the riser from the nearest hydrant.

Dry risers are provided solely for the use of the fire brigade and are therefore not installed for first aid use by the building occupants in the same way as a hose reel. The dry riser is really an extension of the fire brigade hose

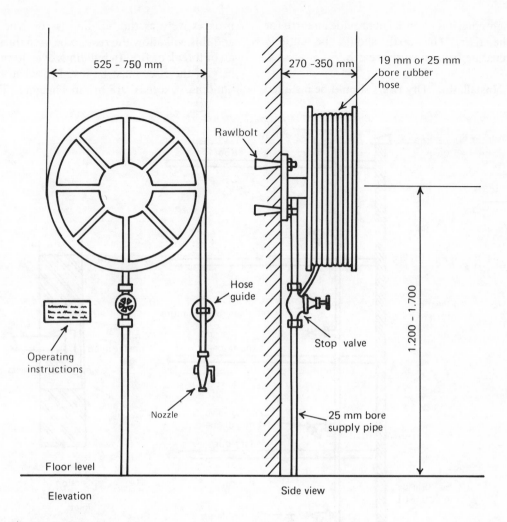

Fig. 4.3 Detail of fixed hose reel.

and avoids the necessity of running long lengths of canvas hose up the staircase of a building from ground level.

Positioning. Dry risers are usually positioned in a ventilated lobby approach to the staircase and this enables the fire brigade to connect their hose pipes to a landing valve in a smoke-free area.

Diameter of Riser. In buildings up to 45 m in height and where there is one 64 mm landing valve on each floor, the internal diameter of the riser should be 100 mm. In buildings between 45 m and 60 m in height, the internal diameter of the riser should be 150 mm. A 150 mm riser is also required for

any building having two 64 mm landing valves on each floor. It is not permissible to install a dry riser in a building above 60 m in height and a wet riser is required for these buildings.

Number of Risers. Dry rising mains should be disposed so that no part of the floor is more than 61 m from a landing valve, the distance measured along a route suitable for a hose line including any distance up and down a stairway. Outlets should be provided for every 930 m² of floor area from the ground level to the roof.

Earthing. In order to prevent the risk of electric shock and damage to the riser by lightning, it is essential to provide an earth for the riser. The earth should be entirely separate from any other earth.

Installation. Dry risers should be installed progressively as the building is constructed and this will allow the riser to be used during an outbreak of fire. In buildings over 30 m in height, the riser must be installed when the building exceeds 18 m in height. The

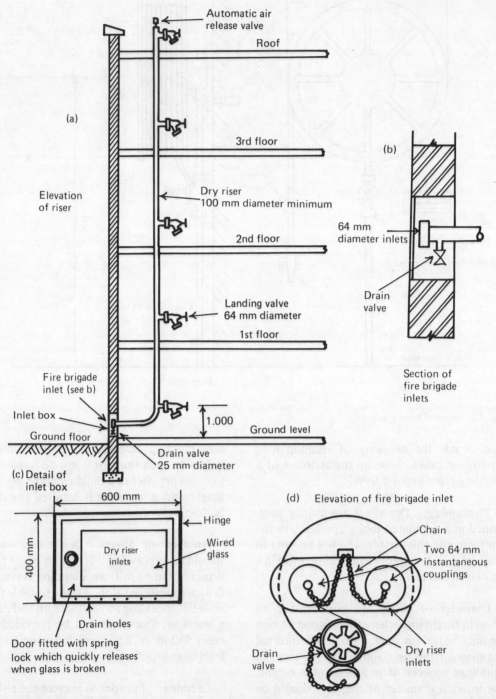

Fig. 4.4 Details of dry riser.

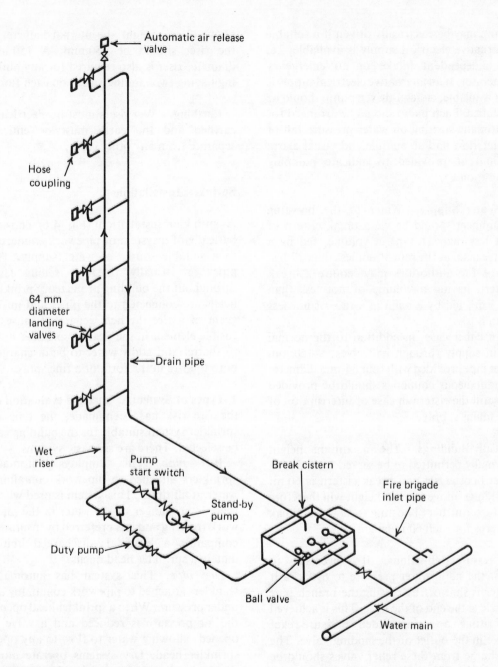

Automatic air release valve

Hose coupling

64 mm diameter landing valves

Drain pipe

Wet riser

Pump start switch

Stand-by pump

Duty pump

Break cistern

Fire brigade inlet pipe

Ball valve

Water main

Fig. 4.5 Installation of a wet riser.

completed installation should be tested and approved by the local fire and water authorities. Fig. 4.4 shows the installation of a dry riser.

Wet Riser (Fig. 4.5)

These are permanently connected to a water supply capable of providing a running pressure of not less than 410 kPa at the top outlet. The maximum running pressure permitted with one outlet in operation is 520 kPa. To maintain the above running pressure it is necessary to employ boosting equipment capable of delivering 23 l/s. Duplicate pumping equipment is required and the stand-by pump may be electrically driven if a suitable alternative electrical supply is available, i.e. an independent intake or an emergency generator. If an alternative electrical supply is not available, a diesel-driven pump should be installed. Each pump should be arranged for automatic starting on water pressure fall or water flow and an audible and visual alarm should be provided to indicate pumping operation.

Water Supply. Water to the boosting equipment should be via a break cistern of not less than 11.4 m³ in volume, fed by a water main at the rate of not less than 27 l/s. Some fire authorities may require a break cistern having a volume of not less than 45.5 m³, fed by a main at a rate of not less than 8 l/s.

In either case, in addition to the normal main supply through ballvalves, a 150 mm inlet pipe provided with four 64 mm diameter instantaneous couplings should be provided to refill the cistern in case of interruption of the mains supply.

Tall Buildings. The maximum height normally permitted to be served from a low-level booster set and break cistern is 60 m. Buildings above 60 m in height will therefore require further boosting sets and break cisterns for each 60 m height.

Pressure Reductions. It is necessary to limit the static water pressure to 690 kPa if water is shut off by closing the branch pipe nozzle at the end of the hose. This is achieved by fitting a spring-loaded pressure-relief valve in the outlet of the landing valve. The discharge from these relief valves should be piped via a 50 mm diameter connection to a 76 mm or 100 mm diameter drain pipe, taken vertically down the building along side the wet riser and discharged over the break cistern.

Diameter of Riser. In buildings up to 45 m in height and where there is one landing valve on each floor, the internal diameter of the riser should be 100 mm. In buildings above 45 m in height, the internal diameter of the riser should be 150 mm. A 150 mm diameter riser is also required for any building having two landing valves on each floor.

Earthing. Wet risers must be electrically earthed and the earth must be entirely separate from any other earth.

Sprinkler Installations

A sprinkler installation (Fig. 4.6) consists basically of a system of pipework connected to a suitable source of water supply. The pipes are usually fixed at ceiling level throughout the building protected. Sprinkler heads are connected to the pipes and in the event of a fire the heat generated causes a fusible element in the nearest sprinkler head to shatter and allow water to be discharged onto the fire in the form of a fine spray.

Types of System. Once the evaluation of the fire risk has been made, the type of sprinkler system suitable for the building can be selected. There are six basic systems.

Wet pipe. This employs automatic sprinklers attached to pipework containing water at all times. This system is used where there is no danger of the water in the pipework freezing and it is preferred by insurance companies as water is discharged immediately a sprinkler head opens.

Dry pipe. This system has automatic spinklers attached to pipework containing air under pressure. When a sprinkler head opens, the air pressure is reduced and a valve is opened, allowing water to flow to any open sprinkler head. Dry systems operate more slowly than wet systems and are more expensive to install and maintain. For these reasons they are normally installed only where there would be a risk of water in the pipework freezing.

Alternate wet and dry. This system is used in unheated buildings and operates as a wet system during the summer months. When winter approaches, the pipework is drained of water and filled with compressed air so

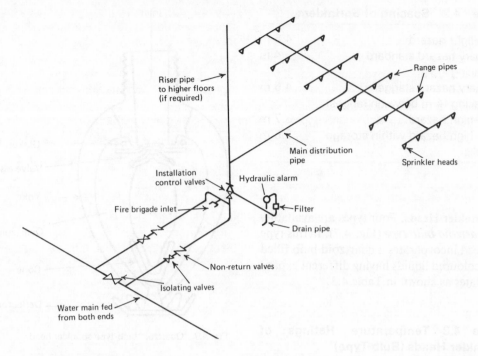

Riser pipe
to higher floors
(if required)

Range pipes

Main distribution
pipe

Sprinkler heads

Installation
control valves

Hydraulic alarm

Fire brigade inlet

Filter

Drain pipe

Non-return valves

Isolating valves

Water main fed
from both ends

Fig. 4.6 Typical wet-pipe sprinkler installation.

that it operates as a dry system during the winter months.

Pre-action systems. These are designed primarily to counteract the operational delay in the dry pipe system and also to eliminate the risk of water damage resulting from accidental damage to sprinkler heads or piping. In the pre-action system, the water supply valve is actuated independently of the opening of the sprinkler heads. The water supply valve is instead opened by the operation of an automatic fire detection system and not by the opening of the sprinkler heads.

Deluge system. The purpose of the deluge system is to deliver the maximum amount of water in the minimum time. The system allows water to cover an entire fire area by admitting water to sprinkler heads or spray nozzles which are open at all times. By using automatic fire detection devices, it is possible to apply water to a fire more quickly than with systems that depend on the opening of sprinkler heads. The system is suitable for extra hazard occupancies in which flammable liquids are stored or handled and where there is a risk that fire may flash ahead of the operation of conventional sprinklers.

Fire-cycle system. In its initial operation, this system is the same as the pre-action system. It has, however, the additional ability to cycle on and off while controlling the fire and to shut itself off when the fire has been extinguished. The system therefore drastically reduces water damage and the on-and-off operation also permits sprinkler heads to be replaced without the necessity of shutting off the main supply valve.

Sprinkler Coverage. The maximum floor area covered by a single sprinkler depends on the fire hazard classification of the building (see Table 4.1). The maximum spacing of sprinklers is as shown in Table 4.2.

Table 4.1 Fire Hazard Classification

Extra-light hazard	21 m²
Ordinary hazard	12 m²
Extra-high hazard	9 m²
Extra-high hazard within storage racks	7.5 m²

Table 4.2 Spacing of Sprinklers

Extra-light hazard	4.6 m
Ordinary hazard standard spacing	4 m
Ordinary hazard staggered spacing (4 m between ranges)	4.6 m
Extra-high hazard	3.7 m
Extra-high hazard within storage racks	2.5 m

Sprinkler Heads. Four types are available.
Quartzoid bulb type (Fig. 4.7). In this type the head incorporates a quartzoid bulb filled with coloured liquids having different expansion rates as shown in Table 4.3.

Table 4.3 Temperature Ratings of Sprinkler Heads (Bulb Type)

Bulb rating (°C)	Colour of bulb liquid
57	Orange
68	Red
79	Yellow
93	Green
141	Blue
182	Mauve
227/288	Black

Side-wall type. This is the same as the quartzoid bulb type but is provided with a deflector at its base so that the spray of water is thrown to one side. They are designed for use at the side of corridors or rooms.
Soldered strut type. The soldered strut consists of three bronze plates joined together by a special low-melting-point solder. These plates hold a glass valve in position against an inlet orifice in a flexible diaphragm and seal the water outlet. When the solder is melted by heat from a fire, the plates fall apart releasing the glass valve and allowing water to spray over the fire.
Duraspeed type. This is an improved version of the simple soldered strut type. The

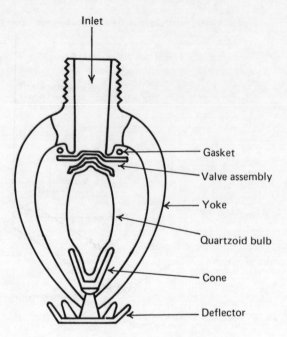

Fig. 4.7 Quartzoid bulb-type sprinkler head.

solder joining together the struts is almost enclosed inside a special element and this reduces the risk of premature operation of the head due to corrosion of the solder. A protective film is applied to the soldered element as a further precaution against atmospheric corrosion.

Sprinkler heads are manufactured with nominal orifice diameters to suit the respective fire hazard classification as shown in Table 4.4.

Table 4.4 Nominal Orifice Diameters of Sprinkler Heads

N.B. Sprinkler heads must never be painted.

Nominal diameter (mm)	Hazard classification
10	Extra light
15	Ordinary
20	Extra high

38

Water Supply for Fire-Control Systems

Water supply from the main, either direct or via a break cistern, and from a high-level storage cistern have been dealt with in the previous descriptions. Other water supply sources that may be suitable are:

1. supply from main and pneumatic cylinder as described in Chapter 3;
2. elevated private reservoir;
3. river or canal.

For the highest insurance rebate, two distinct supplies of water are usually required, i.e. a break cistern supplied from the main and an elevated private reservoir.

10. State the hydraulic requirements of a wet riser.
11. Describe the following sprinkler installations:
    (a) wet pipe,
    (b) dry pipe,
    (c) alternate wet and dry pipe,
    (d) pre-action,
    (e) deluge,
    (f) fire cycle.
12. Sketch a section through the following types of sprinkler head and describe their operation:
    (a) quartzoid bulb type,
    (b) soldered strut type.
13. State the maximum spacings of sprinkler heads.
14. State the various water supply sources that may be used for fire-control systems and the type of supply that will allow the highest insurance rebate.

Exercises

1. Sketch and describe the operation of a hose reel installation for buildings where
   (a) the water mains pressure is sufficient,
   (b) the water mains pressure is insufficient.
2. State the hydraulic requirements for a hose reel installation.
3. Sketch a hose reel installation supplied from a high-level break cistern.
4. State the positions in a building where hose reels should be sited.
5. Sketch and describe the operation of a dry riser for a 10-storey building.
6. State the purpose of a dry riser.
7. State the diameters of dry risers and the diameters of the fire brigade inlets.
8. State the purpose of a wet riser.
9. Sketch and describe the operation of a wet riser for a 25-storey building.

Chapter 5
Sanitary Appliances, Flushing Arrangements, Sanitary Accommodation

Materials Used for Sanitary Appliances

The materials used for sanitary appliances must be durable and easily cleaned and have nonabsorbent surfaces.

Ceramics. The strength and density of these appliances depends on the clay mixture from which they are made and the temperatures at which they are fired. The following ceramic appliances are used:
1. *Vitreous china*. Appliances made from this material are light in weight and nonporous thoughout their thickness. It is generally acknowledged that the material produces the highest quality equipment and a variety of colours can now be obtained.
2. *Stoneware*. This material is normally used for drainage fittings, but it is also used for sinks and washing troughs. The material is very strong and like vitreous china is nonporous throughout its thickness, even when unglazed.
3. *Glazed fireclay*. The material is very strong and appliances such as sinks, urinals and WC pans made from glazed fireclay are often installed in factories and schools where they will withstand rough usage. The material is porous below the glazing and therefore, if the glazing is chipped, the material will absorb water.

Other materials.
1. *Pressed metal*. Mild steel, stainless

steel, and monel metal are moulded by a press to form a one-piece appliance. Mild steel is finished in vitreous enamel on the internal surfaces and a variety of colours can now be obtained.

Galvanised mild steel troughs and sinks are also manufactured where low cost is required. Stainless steel produces a good polished surface which is very resistant to hard wear. The material is used for sinks, WC pans, urinals and drinking fountains. Monel metal has many of the properties of stainless steel and is usually cheaper. It is normally used for sinks.
2. *Acrylic plastics*. The material produces appliances that are very light in weight and cheap to manufacture. A gloss finish can now be obtained, but this must be carefully cleaned to avoid scratching. The plastic becomes soft when heated, so the appliances must always be exposed to cold water before being subjected to hot water, or mixing taps should be used. Baths must be provided with supports, which usually consist of timber on metal cradles. A variety of colours can be obtained.
3. *Glass-reinforced polyester*. Appliances made from this material are much stronger than those made from acrylic plastics, but they are more expensive. Baths and shower trays are

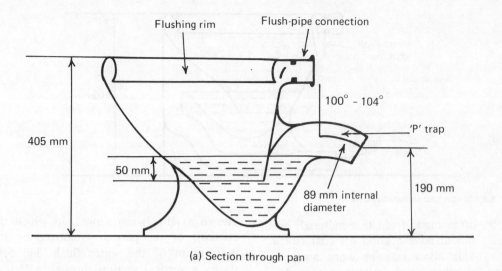

Flushing rim   Flush-pipe connection

405 mm

100° – 104°

'P' trap

50 mm

89 mm internal
diameter

190 mm

(a) Section through pan

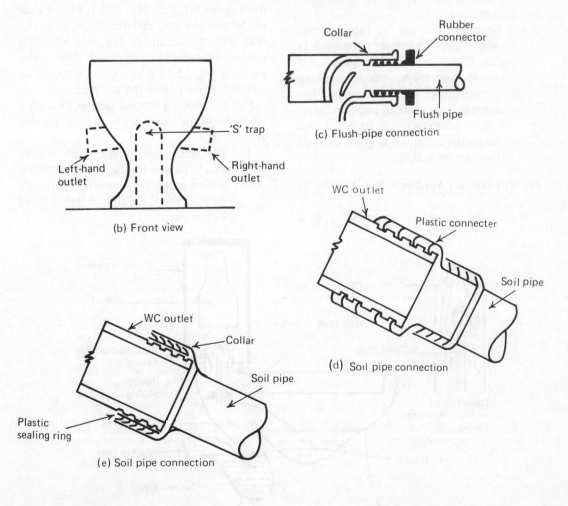

Collar

Rubber
connector

Flush pipe

(c) Flush-pipe connection

'S' trap

Left-hand
outlet

Right-hand
outlet

(b) Front view

WC outlet

Plastic connecter

Soil pipe

(d) Soil pipe connection

WC outlet

Collar

Soil pipe

Plastic
sealing ring

(e) Soil pipe connection

Fig. 5.1 Wash-down WC.

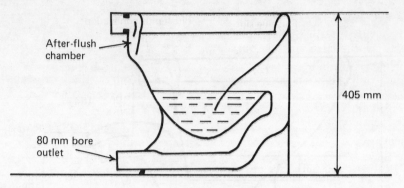

Fig. 5.2 Single-trap siphonic WC pan.

often made from this material, and should have a good gel coat finish. Thin coats may be worn away by cleaning, which could expose the glass fibre.

4. *Terrazzo*. The use of this material enables large appliances to be made in situ. The material is sometimes used for shower trays, baths, washing fountains and washing troughs. The material is very strong and can therefore stand up to rough usage; it also allows the architect a great deal of flexibility of design.

Types of Sanitary Appliance: Soil Appliances

WCs. The most widely used type is

known as the '*wash down*', in which the contents of the pan are removed by the momentum of the water flush. Fig. 5.1a shows a vertical section through a P trap wash-down WC pan. Fig. 5.1b is a front view of the pan showing different types of outlet that may be obtained. Fig. 5.1c shows a method of connecting the flush pipe to the pan and Figs. 5.1d and e show methods of connecting the pan to the soil pipe.

Fig. 5.2 shows a vertical section through a single-trap-type *siphonic* pan, which operates as follows:

1. The pan receives the flush water, which passes through the restricted section and is momentarily restricted: the long leg of the siphon is thus filled with water.

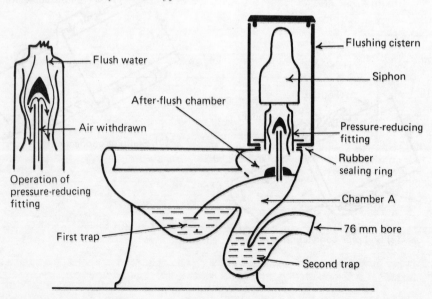

Fig. 5.3 Two-trap-type siphonic WC pan.

2. Siphonic action is set up and the atmospheric pressure acting upon the water in the pan forces out the contents of the pan.
3. Water from the after flush chamber refills the pan.

Fig. 5.3 shows a vertical section through a two-trap-type WC pan, which operates as follows:

1. The cistern is flushed and water passes through the pressure-reducing fitting, A, which draws air through the tube from the space between the two water seals, B.
2. Siphonic action is set up and the atmospheric pressure acting on the

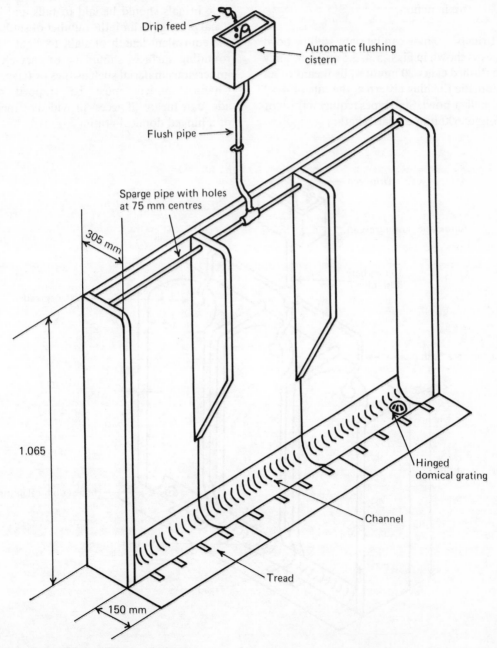

Fig. 5.4 Glazed fireclay slab-type urinal range.

water in the first seal forces out the contents of the pan. At the same time, the flush water flows into the pan to wash the sides clean.

3. Water contained in the after flush chamber flows into the pan to refill the seal. Siphonic WCs are more expensive than the wash-down type, but they are usually more silent and positive in action.

Urinals. These can be slab, stall or bowl types as shown in Figs. 5.4, 5.5 and 5.6. They are flushed every 20 minutes, by means of an automatic flushing cistern at the rate of 4.5 l per stall or bowl. Slab types require a flush of 4.5 l per 600 mm of slab length.

In order to conserve water, the valve supplying the automatic flushing cistern may be a hydraulically operated type, so that the valve is shut off automatically during the hours when the building is unoccupied. Local water authorities usually require water supplying automatic flushing cisterns to be metered.

Channels required for the stall and slab-type urinals should be laid to falls and it is usual practice to limit the number of stalls, or the equivalent length of slab, to eight. Surrounding surfaces should be of easy-clean, impervious material such as tiles or terrazzo. Channel outlets must be trapped and adequate means of access provided by means of a hinged domical grating.

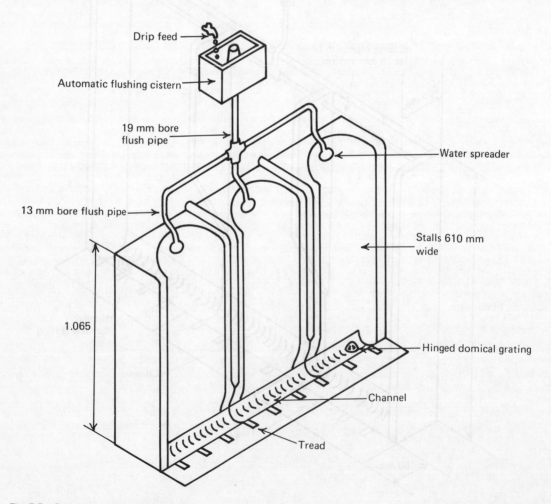

Fig. 5.5 Ceramic stall-type urinal range.

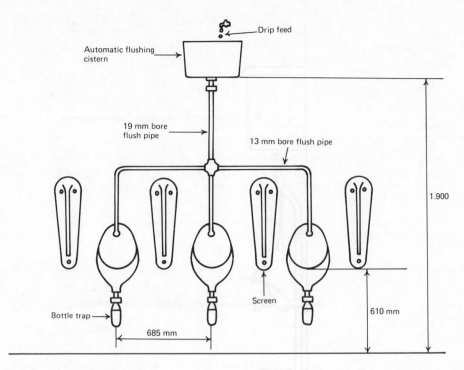

Labels on figure:
- Drip feed
- Automatic flushing cistern
- 19 mm bore flush pipe
- 13 mm bore flush pipe
- 1.900
- 610 mm
- Screen
- Bottle trap
- 685 mm

Fig. 5.6   Ceramic bowl-type urinals.

Unless a suspended ceiling is provided for services, on upper floors, the depth of the trap accommodation is often provided by laying the channels about 150 mm above the floor level on which the urinals are fitted.

Slop Sinks or Hoppers. These are installed in hospitals, colleges, hotels and schools for the disposal of slops. A hinged brass or stainless steel grating is provided to make it easy to fill a bucket, and hot and cold water systems, in the form of antisplash taps, are fitted above bucket height. Fig. 5.7 shows the installation of a slop sink or hopper and Fig. 5.8 the installation of a combined wash-up sink and slop sink.

Bed-Pan Washers.   These are installed in hospital sluice rooms and are provided with a bed pan and urine bottle jet. Since the jet is below the flooding level of the appliance, there is a danger of pollution of water by back siphonage as described in Chapter 2. The jet should therefore be supplied with water from a cold water storage cistern through a separate pipe.

Fig. 5.9 shows the installation of a combined bed-pan washer and wash-up sink.

Types of Sanitary Appliance: Waste Appliances

Washbasins. Many washbasin designs are available, ranging from large hair-dressers' shampoo basins to small hand-rinse basins for use in cloakrooms, where space is limited. Basins can be obtained to fit into the corner of a room, or fitted flush to a vanity unit for a bedroom.

The fittings are usually manufactured for hot and cold pillar taps, but patterns can be obtained having only one hole for a spray mixer tap and these types do not require an overflow or a plug. They also use less water and save fuel. With the usual type of basin, the overflow takes the form of a weir, over which the overflow water passes down a channel formed in the basin to a slot in the waste fitting. Because these channels can be easily fouled, hospital basins are often provided with a stand pipe which also acts as

45

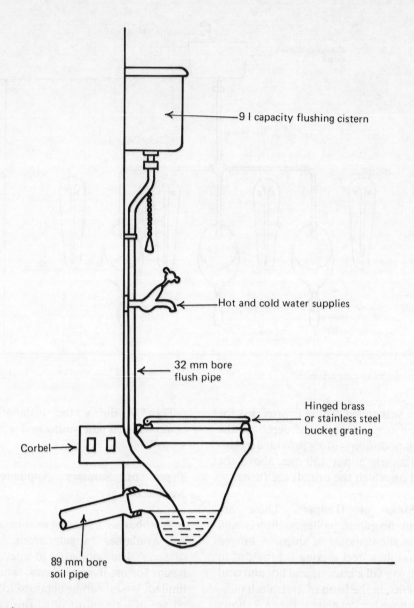

9 l capacity flushing cistern

Hot and cold water supplies

32 mm bore
flush pipe

Hinged brass
or stainless steel
bucket grating

Corbel

89 mm bore
soil pipe

Fig. 5.7 Slop hopper.

a plug. This type of overflow can easily be removed and sterilised.

Most designs of washbasin include a moulded soap tray at the back or at either side. Basins are fixed and supported by cantilever brackets, brackets and legs, or pedestals. Fig. 5.10 shows a washbasin supported by a pedestal. The pedestal conceals the pipework but it makes it more difficult to clean behind. Some basins have ceramic lugs which can be built into a load-bearing wall; these are often called corbel types.

Ablution or Washing Fountains (Fig. 5.11). These may be used in schools and factories as an alternative to washbasins. They provide a very hygienic means of hand washing because, when in use, clean water is continually flowing. They also take up less space than an equivalent number of wash-basins. Up to 12 people may use the fountain

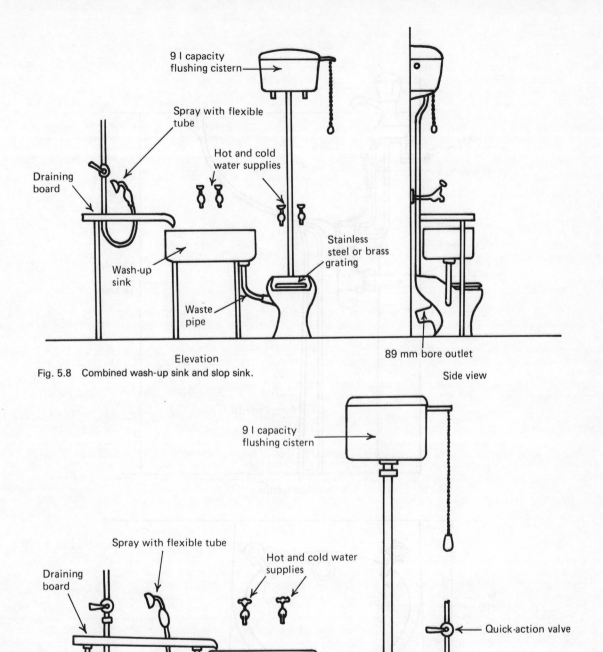

9 l capacity
flushing cistern

Spray with flexible
tube

Hot and cold
water supplies

Draining
board

Wash-up
sink

Waste
pipe

Stainless
steel or brass
grating

89 mm bore outlet

Elevation

Fig. 5.8   Combined wash-up sink and slop sink.

Side view

9 l capacity
flushing cistern

Spray with flexible tube

Hot and cold water
supplies

Draining
board

Quick-action valve

Nozzle

Wash-up
sink

Bed pan washer
supported by corbel
with 89 mm bore outlet

Waste pipe

Fig. 5.9   Bed-pan washer and sink combination.

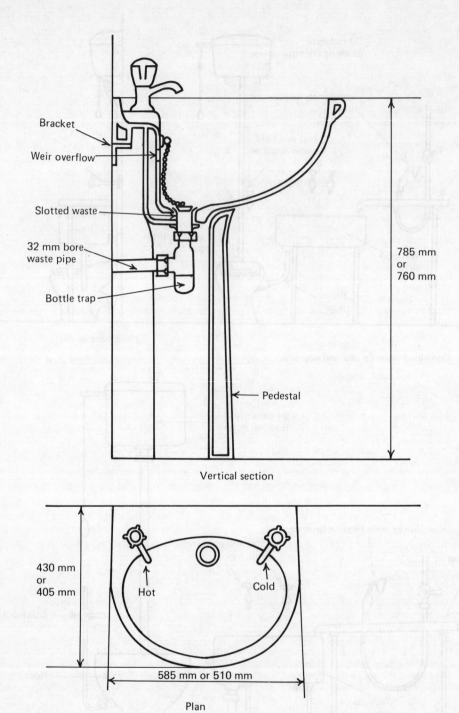

Bracket

Weir overflow

Slotted waste

32 mm bore
waste pipe

Bottle trap

Pedestal

785 mm
or
760 mm

Vertical section

430 mm
or
405 mm

Hot

Cold

585 mm or 510 mm

Plan

Fig. 5.10 Washbasin with pedestal.

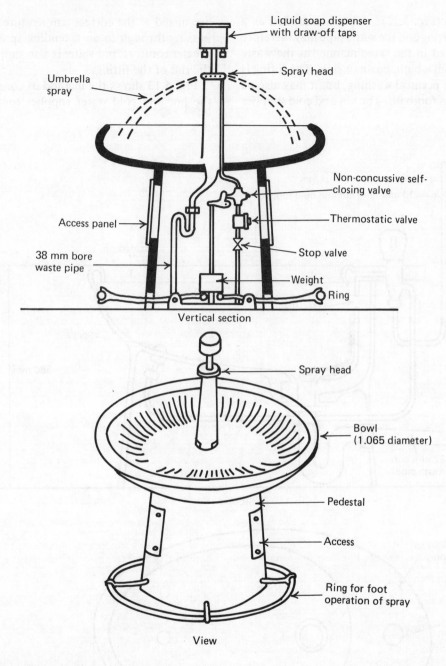

Liquid soap dispenser
with draw-off taps

Spray head

Umbrella
spray

Non-concussive self-
closing valve

Thermostatic valve

Access panel

Stop valve

38 mm bore
waste pipe

Weight

Ring

Vertical section

Spray head

Bowl
(1.065 diameter)

Pedestal

Access

Ring for foot
operation of spray

View

Fig. 5.11 Ablution fountain.

and the flow of hot water to the umbrella spray is controlled by a steel ring. When the ring is depressed by the foot, an iron weight at the end of a lever is lifted and at the same time a steel rod connected to a weight opens a non-concussive self-closing valve. Hot and cold water then flows through a thermostatically controlled valve and passes through a pipe to a spray head. The spray head is designed to provide an umbrella-shaped spray of water at the correct temperature for hand washing.

Bidet (Fig. 5.12). This is classified as a waste fitting and the waste pipe may therefore be treated in the same manner as the waste pipe from a bath, basin or sink. The fitting is used for perineal washing, but it may also be used as a footbath. The hot and cold supplies are mixed at the correct temperature before passing through to an ascending spray. For greater comfort, hot water is also supplied to the rim of the fitting.

Fig. 5.13 shows the method of connecting the hot and cold water supplies to a bidet

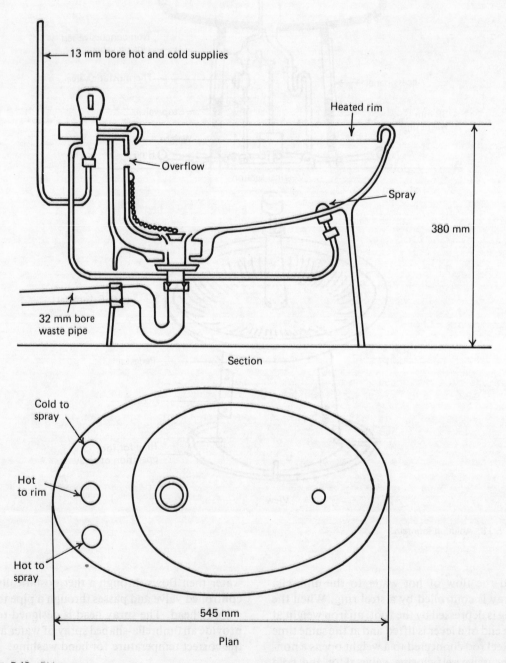

13 mm bore hot and cold supplies

Heated rim

Overflow

Spray

380 mm

32 mm bore
waste pipe

Section

Cold to
spray

Hot
to rim

Hot to
spray

545 mm

Plan

Fig. 5.12   Bidet.

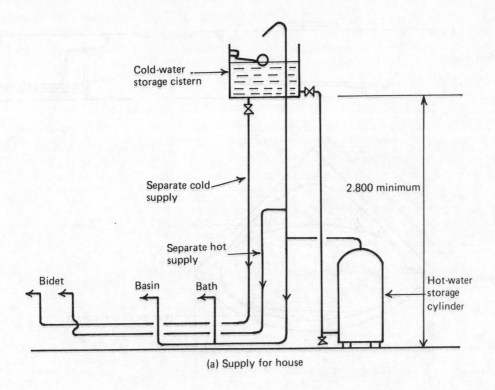

(a) Supply for house

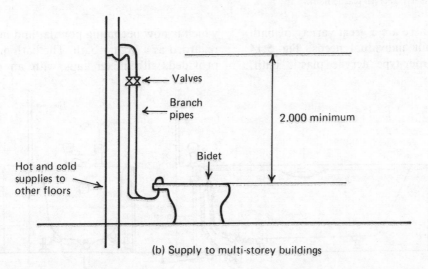

(b) Supply to multi-storey buildings

Fig. 5.13   Water supply to bidets.

fitted in (a) a house and (b) a multi-storey building. Because of the low-level spray inlet, it is important that foul water from the bidet is prevented from siphoning or flowing back into the supplies to other fittings.

Separate hot and cold water supplies are therefore required, as shown in the diagram for supply for a house. When bidets are fitted on each floor of a multi-storey building, the separate hot and cold water supplies must be connected to the vertical mains, at a height of at least 2 m above the top level of the bidet. This is a precaution taken to reduce the risk of water from the bidet being siphoned back into the main pipes.

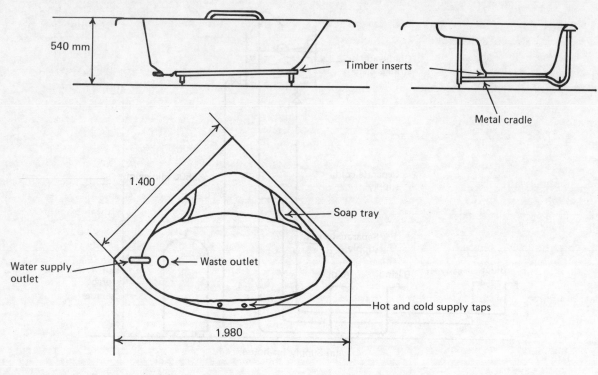

Fig. 5.14   Corner-type acrylic plastic bath.

Baths.   There are a great variety of bath shapes to suit individual needs. Fig. 5.14 shows a corner-type acrylic plastic bath, which is now becoming popular and may be regarded as a luxury bath. The bath may be provided with mixer taps with an outlet

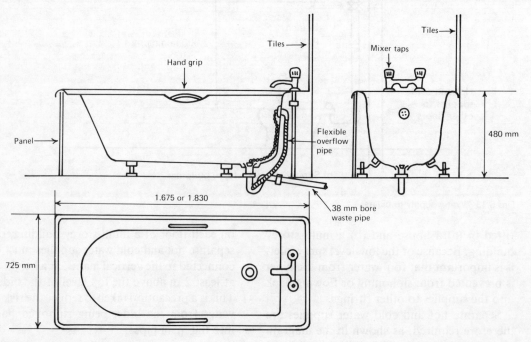

Fig. 5.15   Enamelled cast-iron bath.

placed at one corner, to make it easier to enter or leave the bath. The bath is provided with adequate supports in the form of a metal cradle with timber inserts.

Fig. 5.15 shows a traditional enamelled cast-iron bath with hand grips and mixer taps. If required, the mixer taps may also be provided with a flexible connection for a spray head, which may be used for a shower. Cast-iron baths are very heavy and require adequate supports which are provided by the manufacturer in the form of adjustable feet.

Fig. 5.16 shows a detail of a sitz bath which is deep and incorporates a seat. It may be used where space is limited and is also suitable for

old people since the user may maintain a sitting position.

Showers. These use only about a third of the water required for a bath and therefore economise in water consumption and also fuel. They are much quicker to use than a bath and take up less space: many people however prefer a good 'soak' in a bath. Two types of shower head available are:

1. the traditional rose type, which discharges water through a perforated disc, and which is fitted at high level, and

2. an adjustable umbrella-spray type,

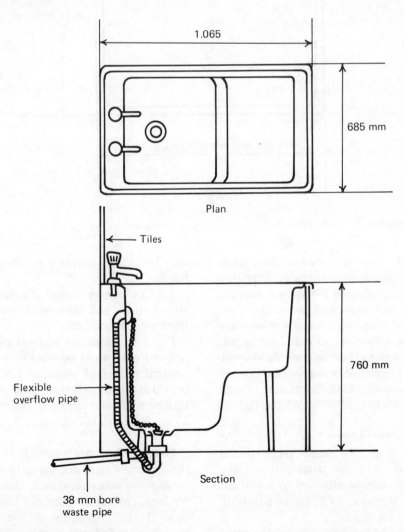

Fig. 5.16 Enamelled cast-iron sitz bath.

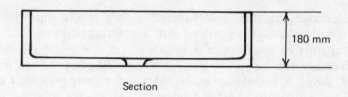

Section

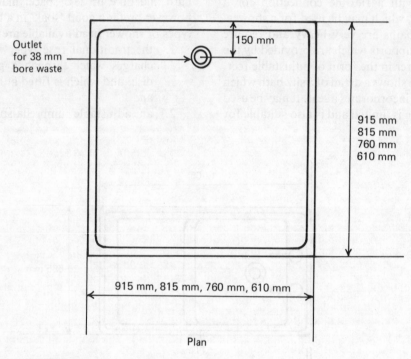

Outlet
for 38 mm
bore waste

150 mm

180 mm

915 mm
815 mm
760 mm
610 mm

915 mm, 815 mm, 760 mm, 610 mm

Plan

Fig. 5.17   Glazed fireclay shower tray.

which is usually fitted at chest level. This type is sometimes preferred by women, because the shower may be used without wetting the head.

Hot and cold water supplies to a shower are provided by means of 13 mm bore pipes, which should be as short as possible in order to reduce frictional resistances to the flow of water. A thermostatically controlled mixing valve is recommended to avoid the risk of scalding. The outlet from the mixing valve may be by means of a pipe behind the tiles, or a flexible chromium-plated pipe to the shower head. With the latter method, the shower head may be attached to a vertical bracket which provides a means of adjusting the height of the head.

In schools and factories, one mixing valve

may be used to supply a number of shower heads.

Fig. 5.17 shows a detail of a glazed fireclay shower tray, and glass-reinforced polyester trays are also available.

Fig. 5.18 shows the method of providing hot and cold water supplies to a shower. A minimum head of water of 1 m should be provided at the shower head and manufacturers often specify a maximum head of 20 m above the mixing valve.

Sinks.   These are available in almost all the materials used for sanitary appliances. Sinks serve many functions, including dishwashing, clothes washing and food preparation. There are many different designs, including the following.

54

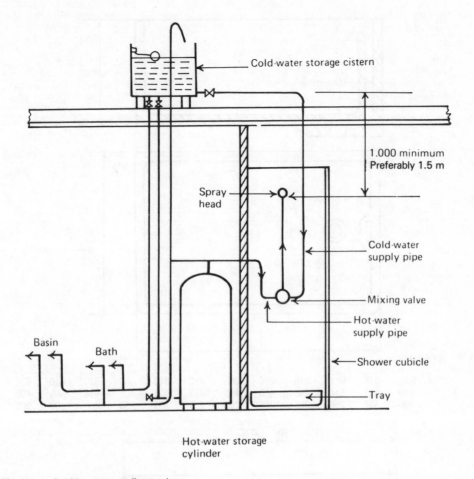

Cold-water storage cistern

1.000 minimum
Preferably 1.5 m

Spray head

Cold-water supply pipe

Mixing valve

Hot-water supply pipe

Shower cubicle

Tray

Basin

Bath

Hot-water storage cylinder

Fig. 5.18 Hot and cold water supplies to shower.

Belfast sinks (Fig. 5.19). These are made from glazed fireclay and are supported on cantilever brackets or brackets and legs. The sinks have an integral weir overflow.

London sinks. These are very similar to Belfast sinks, but are not provided with an integral weir overflow.

Combination sinks (Fig. 5.20). These are again made from glazed fireclay and are a Belfast-type sink with an integral weir overflow. The sink is provided with an integral draining surface.

Cleaners' sink (Fig. 5.21). These are often fitted in schools, factories and hospitals and may be installed inside the cleaners' cubicle or the sanitary accommodation room. They are provided with a hinged brass or stainless steel grating, which is used to support a bucket whilst it is being filled.

Stainless steel sink (Fig. 5.22). These are very popular and are used in all types of building. For domestic buildings, the sink may have single or double drainers, also single or double bowls. Large sinks may be obtained for canteen kitchens, and these may have three bowls, which makes them very useful for food preparation and dish washing.

Drinking Fountains

These are required for offices, schools, colleges, hospitals, factories and most business premises. Several patterns are available in glazed fireclay or stainless steel and the appliance may be a pedestal type or corbelled from a loadbearing wall as shown in Fig. 5.23.

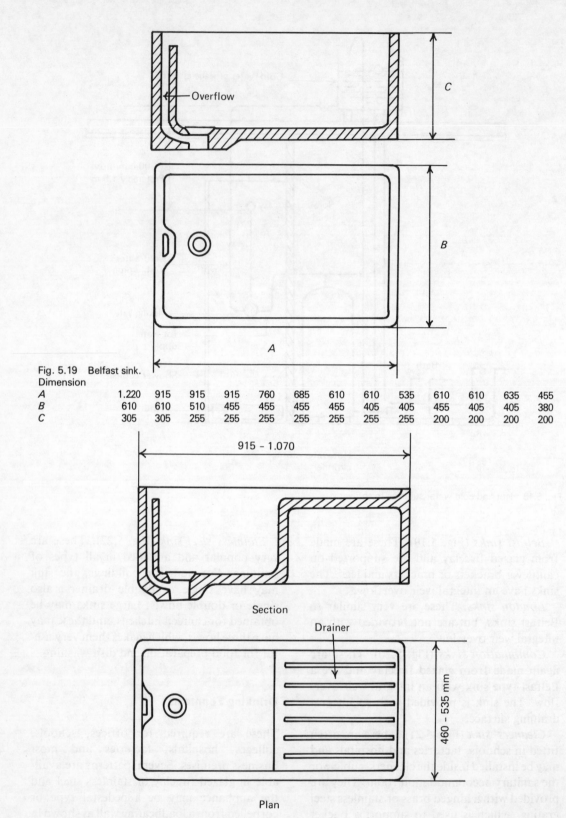

Overflow

Fig. 5.19   Belfast sink.
Dimension

A	1.220	915	915	915	760	685	610	610	535	610	610	635	455
B	610	610	510	455	455	455	455	405	405	455	405	405	380
C	305	305	255	255	255	255	255	255	255	200	200	200	200

915 – 1.070

Section

Drainer

460 – 535 mm

Plan

Fig. 5.20   Glazed fireclay combination sink.

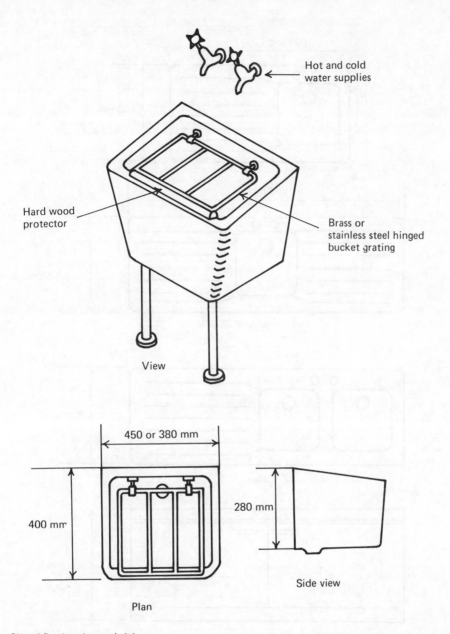

Hot and cold
water supplies

Hard wood
protector

Brass or
stainless steel hinged
bucket grating

View

450 or 380 mm

400 mm

280 mm

Plan

Side view

Fig. 5.21 Glazed fireclay cleaners' sink.

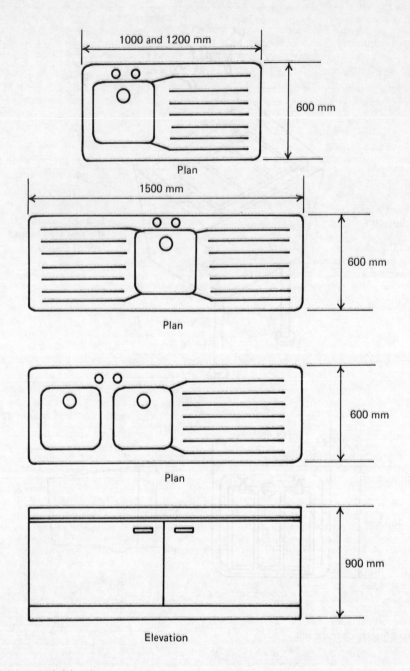

Fig. 5.22 Stainless steel sink units.

58

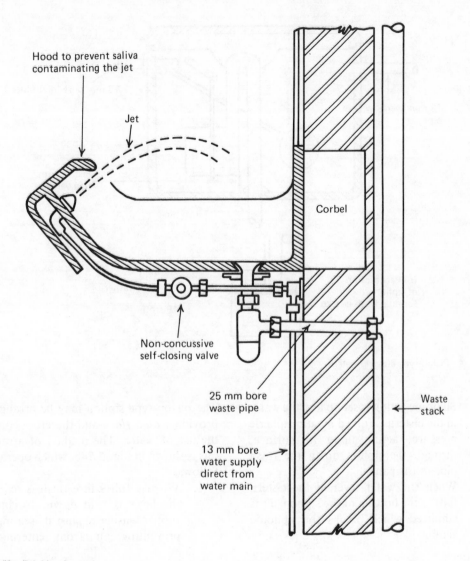

Hood to prevent saliva
contaminating the jet

Jet

Corbel

Non-concussive
self-closing valve

25 mm bore
waste pipe

13 mm bore
water supply
direct from
water main

Waste
stack

Fig. 5.23 Drinking fountain.

It is important that the water outlet is hooded, to avoid water from the mouth falling on the jet. The water suply to the fountain should come directly from the water main and precautions should be taken against the water being heated or contaminated. The water supply should therefore not be placed near to a heating or hot water pipe and it is in any case good practice to lag the pipe to prevent heat gains from other sources. If required, the water may be chilled by means of a refrigerator unit.

Flushing Cisterns

There are two types available.

Piston types, as show in Fig. 5.24, operate as follows.
1. The lever outside the cistern is depressed, which causes the piston inside the siphon to be raised.
2. The flap valve closes and water above the valve is ejected through the siphon and down the flush pipe.

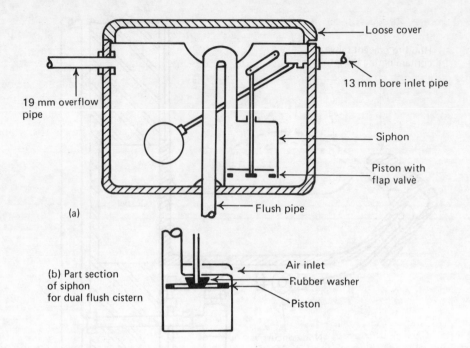

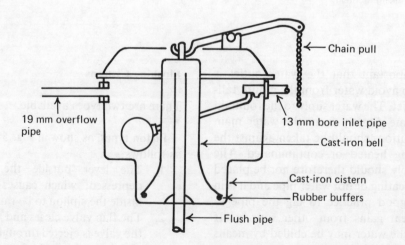

Fig. 5.24   Piston-type flushing cistern.

3. Siphonic action is set up and the water in the cistern is forced by atmospheric pressure, acting upon its surface, through the siphon down the flush pipe to the pan.

4. When the water level in the cistern falls to the bottom of the siphon, air is admitted which breaks the siphonic action.

The piston-type siphon may be arranged to provide a *dual flush* and therefore conserve the use of water. The method of arranging this is shown in Fig. 5.24b, which operates as follows.

1. When a full 9 litre flush is required, the lever is held down, forcing the rubber washer against its seating and preventing air from entering the

Fig. 5.25

siphon, and a full flush is therefore obtained.

2. When only a 4.5 litre flush is required, the lever is released in the normal way and this breaks the siphonic action at a point when the water level reaches the air hole, by allowing air to enter the siphon. Only half a flush is therefore obtained.

Bell types, as shown in Fig. 5.25, operate as follows.

1. The lever is depressed which lifts the cast iron bell.
2. The lever is released and the bell falls, displacing the water below it over the stand pipe.
3. Water falling down the stand pipe displaces the air and a siphonic action

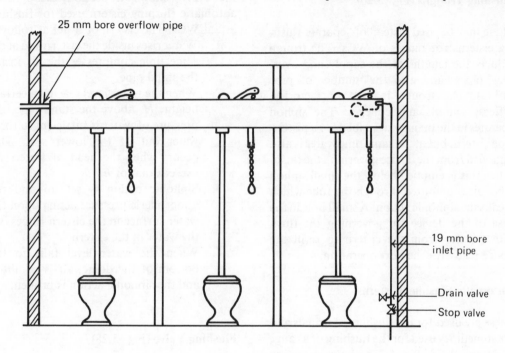

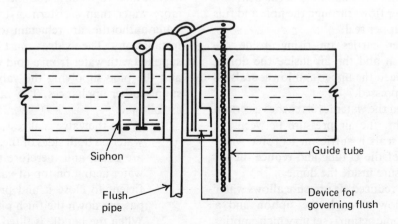

Fig. 5.26 Flushing trough.

4. The atmospheric pressure acting upon the water out of the cistern to flush the pan.
5. When the water level in the cistern falls to the bottom of the bell, air is admitted, which breaks the siphonic action.

Flushing Troughs (Fig. 5.26)

These may be used instead of separate flushing cisterns for ranges of WCs. The trough reduces the lengths of the supply and overflow pipes, and also the number of pipe fittings. The trough is used in factories, offices, schools and colleges. The siphon operates in the manner described for a piston-type cistern, but at the same time water is also siphoned from the device governing the flush. When this is emptied below the small siphon tube, air is admitted through the tube which breaks the siphonic action. A small hole in the base of the device for governing the flush allows water from the cistern to refill the device ready for the next operation.

Automatic Flushing Cisterns

These are used for the flushing of urinals and are sometimes used for the flushing of a range of infant WCs. Fig. 5.27a shows a section through an automatic flushing cistern used for urinals. The cistern operates as follows.
1. Water rises inside the cistern until it reaches the small hole in the bell.
2. Water flows through the hole and fills the upper seal.
3. Water carries on rising inside the cistern and the air inside the dome, between the upper and lower seals, is compressed.
4. When the water level reaches a certain height, the air pressure between the two seals is sufficient to force water out of the U tube and reduce the air pressure inside the dome.
5. This reduced air pressure allows water to flow through the siphon and a siphonic action is set up which empties the cistern.

6. When the water level falls to the bottom of the dome, air is admitted and the siphonic action is broken.
7. When the flush has finished, water from the upper seal is siphoned through the siphon tube into the lower seal and the cistern is then ready for another flushing operation.

Fig. 5.27b shows a section through an automatic flushing cistern used for flushing infant WCs. The cistern operates as follows.
1. Water rises inside the cistern and at the same time compresses the air inside the stand pipe.
2. When the water level reaches a certain height, H, above the stand pipe, the pressure of water is sufficient to force water out of the lower seal. This occurs when the head of water, H, exceeds that of h.
3. Siphonic action is set up and the atmospheric pressure acting upon the water surface in the cistern forces out the water in the cistern.
4. When the water level falls to the bottom of the dome, air is admitted and the siphonic action is broken.

Flushing Valve (Fig. 5.28)

These are sometimes used as an alternative to the flushing cistern or trough for the flushing of WCs. Because of the risk that the valve may be incorrectly adjusted and therefore use more water than a cistern or trough, some water authorities are reluctant to permit this installation. The valve must always be supplied with water from a cold water cistern and the pipe supplying the valve must not supply other appliances. The valve operates as follows.
1. When the valve is at rest, the force of water on both sides of the cup washers are equal and therefore the force of water acting on top of valve A is sufficient to close it and prevent water passing down the flush pipe.
2. When the handle is tilted, the release valve B is also tilted and water escapes

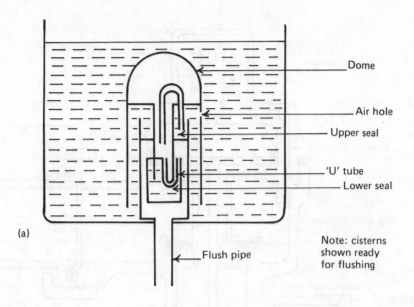

Dome

Air hole

Upper seal

'U' tube

Lower seal

(a)

Note: cisterns
shown ready
for flushing

Flush pipe

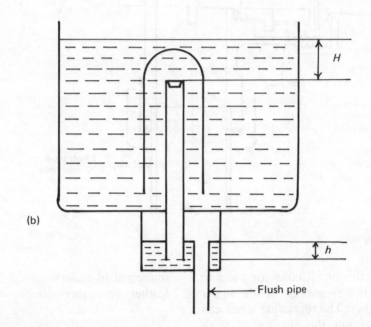

H

(b)

h

Flush pipe

Fig. 5.27 Automatic flushing cistern.

from the upper pressure chamber
down the flush pipe.

3. The force of water acting on the
   underside of the lower cup washer is
   greater than the force acting on the
   upper cup washer and valve A is therefore raised, allowing water to flow
   down the flush pipe.

4. During the flush, water flows through
   the by-pass into the upper pressure
   chamber and this again equalises the
   force of water acting on both sides of
   the cup washers. The force of water
   acting on top of valve A forces it onto
   its seating and stops the flush.

Note: the length of time the flush operates is

63

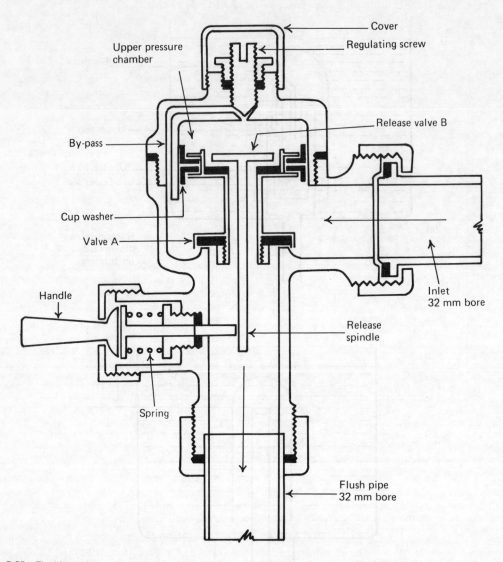

Fig. 5.28 Flushing valve.

determined by the time it takes for water to flow through the by-pass to fill the upper pressure chamber. The regulating screw can be adjusted to suit the amount of flush required. Screwing it down will restrict the by-pass and the time taken to fill the upper pressure chamber is increased, thus increasing the amount of flush.

Fig. 5.29 shows the installation of flushing valves for a range of three WCs. The stop valves shown, connected to the flushing valves, are usually provided by the manufacturer of the fitting. Before installing flushing valves, it is important that they are

immersed in water overnight to soften the leather cup washers.

Waste Disposal Units

These units are designed to dispose of organic food waste in domestic and canteen kitchens. The units must not be used for the disposal of rags, plastics, metal or glass.

Fig. 5.30a shows a domestic-type waste disposal unit which uses a 153 W electric motor. The current used, when the unit is in operation, is 3 A, but the starting current on

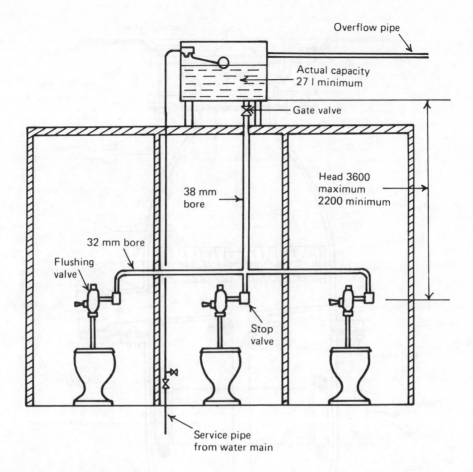

Overflow pipe

Actual capacity
27 l minimum

Gate valve

38 mm
bore

Head 3600
maximum
2200 minimum

32 mm bore

Flushing
valve

Stop
valve

Service pipe
from water main

Fig. 5.29   Installation of flushing valve.

some models can be up to 13 A. The annular velocity of the cutter rotor is about 1450 rev./min. It is essential that three-core electric cable is used, so that the unit can be earthed. The unit should be provided with electricity through a fused control switch, incorporating a pilot light which lights up an indicator lamp when the motor is running. Units for canteen kitchens are large and require electric motors ranging from 230 W to 4.6 kW in power. The unit operates as follows.

1. Cold water is run into the sink and the unit is switched on.
2. Waste food is pushed through the rubber splash guard and is washed down by the water onto the high-speed cutter rotor (the rotor is too low down to risk danger to the user).
3. The waste food is flung by centrifugal force onto the stationary cutting ring, cutting the food into fine particles.
4. The partially liquefied food particles discharge through the waste pipe to a back gully via the discharge chamber.

Note: the outlet of the trap connected to the unit should be well below the inlet to prevent water lying inside the discharge chamber. The units are provided with a thermal overload switch in case of jamming.

Fig. 5.30b shows the waste pipe arrangement for the unit, the trap may be a P type as shown or an S trap.

Sanitary Conveniences

The list of provisions required to comply with the Building Regulations 1985, for entry and performance is as follows.

65

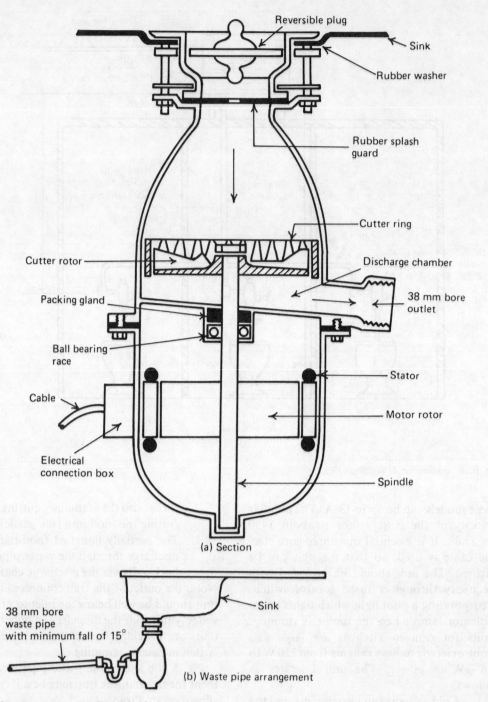

Reversible plug

Sink

Rubber washer

Rubber splash guard

Cutter ring

Cutter rotor

Discharge chamber

38 mm bore outlet

Packing gland

Ball bearing race

Stator

Cable

Motor rotor

Electrical connection box

Spindle

(a) Section

38 mm bore waste pipe with minimum fall of 15°

Sink

(b) Waste pipe arrangement

Fig. 5.30 Domestic waste-disposal unit.

1. Sufficient sanitary conveniences shall be provided which shall be:
   (a) in rooms separated from places where food is stored or prepared and

   (b) designed and installed so as to allow effective cleaning.

Figure 5.31 shows the method of ventilating a sanitary convenience having an external wall and Fig. 5.31 shows the method of separating

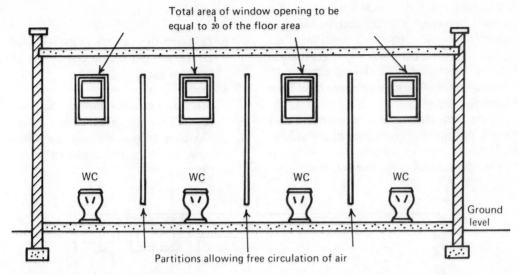

Total area of window opening to be equal to $\frac{1}{20}$ of the floor area

WC   WC   WC   WC

Ground level

Partitions allowing free circulation of air

Fig. 5.31   Ventilation.

a sanitary convenience from a kitchen or workroom. The intervening ventilated space will effectively separate the sanitary convenience from the other rooms. Figures 5.33 and 5.34 show the method of entering a bathroom in a dwelling. If possible it is good practice to separate the bathroom containing a WC from the bedroom by providing an intervening ventilated space, especially if the only WC on the premises is in the bathroom. A landing or corridor will act as a ventilated space.

2. To achieve an acceptable level of performance to reduce the risk to health of persons in the buildings, closets should be provided which are
   (a) in sufficient number and of the appropriate type for the age and sex of the persons using the building, and
   (b) sited, designed and installed so as not to be prejudicial to health.

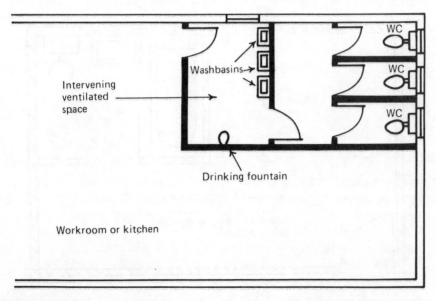

Intervening ventilated space

Washbasins

WC

WC

WC

Drinking fountain

Workroom or kitchen

Fig. 5.32   Method of entry.

67

Ventilation. If mechanical ventilation of a sanitary convenience is required the Building Regulations 1985 require a minimum air change of three per hour. The ventilation may be intermittent but should run for 15 minutes after the convenience has been vacated.

If the convenience has an external wall natural ventilation may be used by a window, skylight or similar means of ventilation which opens directly into the external air.

Exercises

1. Describe the various materials used for sanitary appliances and state the advantages of each.
2. Define
   (a) waste sanitary appliances,
   (b) soil sanitary appliances.
3. Sketch a section through the following siphonic WCs and explain their operation:

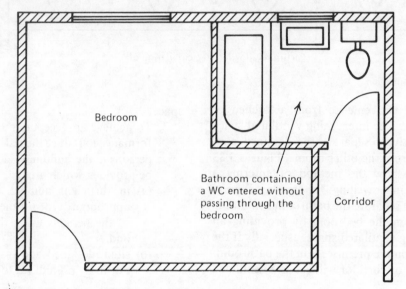

Bedroom

Bathroom containing a WC entered without passing through the bedroom

Corridor

Fig. 5.33 Bathroom containing a WC without another WC within the dwelling.

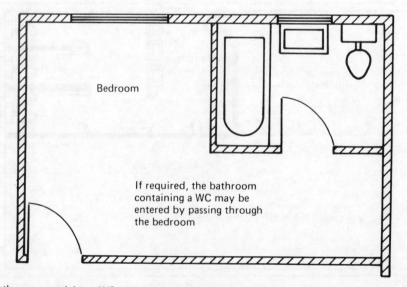

Bedroom

If required, the bathroom containing a WC may be entered by passing through the bedroom

Fig. 5.34 Bathroom containing a WC with another WC within the dwelling.

(a) single-trap type,

(b) two-trap type.

4. Sketch the following types of urinal:

(a) slab type,

(b) stall type,

(c) bowl type.

5. Sketch sections of the following appliances:

(a) slop sink,

(b) bed-pan washer,

(c) washbasin.

6. Sketch sections of the following appliances showing the method of providing hot and cold water supplies and waste to each:

(a) bidet,

(b) washing fountain,

(c) bath.

7. State the advantages of a shower and sketch the method of installing a shower, including the hot and cold water supplies.

8. Describe the following types of sink:

(a) Belfast,

(b) London,

(c) combination,

(d) cleaners,

(e) stainless steel.

9. Sketch and describe the installation of a drinking fountain.

10. Sketch sections and explain the operation of the following types of flushing cistern:

(a) piston type,

(b) bell type,

(c) trough type.

11. Explain how a flushing cistern may be designed to operate on a dual flush principle and state the purpose of this type of cistern.

12. Sketch a section through an automatic flushing cistern and explain its operation.

13. Sketch a section through a flushing valve, explain its operation, and show how flushing valves may be installed for a range of three WCs.

14. Sketch and describe a waste disposal unit, including the safety devices.

15. Explain the Building Regulations (1985) for the provision of entry and ventilation for a kitchen.

16. A new integral sanitary block containing three WCs and four washbasins is to be installed adjacent to a workshop. Sketch a plan of the accommodation showing the method of entry and ventilation.

Chapter 6
Soil and Waste Systems

Definitions

The British Standard Code of Practice for sanitary pipework, 5572:1978 defines the following.

ACCESS COVER: a removable cover on pipes and fittings providing access to the interior of pipework for the purpose of inspection, testing and cleansing.

BRANCH DISCHARGE PIPE: a discharge pipe connecting sanitary appliances to a discharge stack.

BRANCH VENTILATING PIPE: a ventilating pipe connected to a branch discharge pipe.

CRITERION OF SATISFACTORY SERVICE: the percentage of time during which the design discharge flow loading will not be exceeded.

CROWN OF TRAP: the topmost point of the inside of a trap outlet.

DEPTH OF WATER SEAL: the depth of water that would have to be removed from a fully charged trap before air could pass freely through the trap.

DISCHARGE PIPE: a pipe that conveys the discharges from sanitary appliances.

DISCHARGE UNIT: a unit so chosen that the relative load-producing effect of sanitary appliances can be expressed as multiples of that unit. The discharge unit rating of an appliance depends upon its rate and duration of discharge, on the interval between discharges and on the chosen criterion of satisfactory service. It is not a simple multiple of the flow rate.

SIZE: the term used to indicate the nominal internal diameter of the pipes regardless of specific materials and their classification or description.

STACK: a main vertical discharge or ventilating pipe.

TRAP: a fitting or part of an appliance or pipe arranged to retain water or fluid so as to prevent the passage of foul air.

VENTILATING PIPE: a pipe provided to limit the pressure fluctuations within the discharge pipe.

Regulations

The Building Regulations 1985 Part H.1 gives the following requirements for soil pipes, waste pipes and ventilating pipes.

1. *Branch ventilating pipes*
   (a) A branch ventilating pipe should be connected to the discharge pipe within 300 mm of the trap and should not connect to the stack below the 'spillover' level of the highest appliance served.
   (b) Branch ventilating pipes which run direct to the outside air should finish at least 900 mm above any opening into the building nearer than 3 m.
   (c) Branch ventilating pipes to branch pipes serving one appliance should be 25 mm diameter or, where the branch is longer than 15 m or has more than five bends, should be at least 32 mm diameter.

Note: the 'spillover' level of an appliance is the level at which water would flow over when the appliance is overfilled. It is not the level of the weir overflow, see Fig. 6.2.

2. *Access for clearing blockages*
   Rodding points should be provided to give access to any lengths of discharge pipes which cannot be reached by removing traps.

Note: the top of a vent pipe may provide access.

3. *Branch discharge pipes*
   (a) A branch pipe should not discharge into a stack less than 750 mm above

the invert of the tail of the bend at the foot of the stack in a multi-storey building up to five storeys.

(b) A branch pipe serving any ground-floor appliance may discharge direct to a drain or into its own stack. If the building has more than five storeys, ground floor appliances, unless discharging to a gully or drain, should discharge into their own stack. If the building has more than twenty storeys, ground floor appliances, unless discharging into a gully or drain, and first floor appliances, should discharge into their own stacks.

4. *Discharge stacks*

(a) All stacks should discharge to a drain. The bend at the foot of the stack should have as large a radius as possible and at least 200 mm at the centre line.

(b) Stacks should not have offsets in any part carrying foul water (the wet part below the highest branch) and should be run inside the building if it has more than three storeys.

5. *Ventilation of discharge stacks*

(a) To prevent the water seals in the traps from being drawn by pressure which can develop in the system, drainage stacks should be ventilated. However an unvented stack (stub stack) may be used if the stack connects into a ventilated discharge stack or a drain and no branch into the stack is more than 2 m above the invert of the connection or drain and no branch serving a closet is more than 1.5 m from the crown of the closet trap to the invert of the connection or drain (see Fig. 6.26).

(b) Where a stub stack is used there may still be other parts of the system which should be ventilated.

(c) The termination of ventilating pipes open to outside air should be as given in branch ventilating pipe part (b). The termination should be finished with a cage or other cover

which does not restrict the flow of air (see Fig. 6.24).

(d) Discharge stacks may terminate inside a building when fitted with air admittance valves. Where these valves are used they should not adversely affect the ventilation necessary to the below-ground drainage system which is normally provided by the open stacks of the sanitary pipework. Only an air admittance valve which is the subject of a current British Board of Agrément Certificate should be used and the conditions of use should be in accordance with the terms of the certificate.

(e) The size of the part of a stack which serves only for ventilation (i.e., the dry part above the highest branch) may be reduced in one- and two-storey houses, but should be at least 75 mm diameter.

6. *Sizes of discharge stacks*
Stacks serving urinals should not be less than 50 mm diameter and stacks serving closets not less than 75 mm diameter (siphonic WC with 75 mm outlet).

7. *Watertightness*
The installation should be capable of withstanding an air or smoke test of positive pressure of at least 38 mm water gauge for at least three minutes. During this time every trap should maintain a water seal of at least 25 mm.

8. *Ventilation stacks*

(a) A dry stack may provide ventilation for branch ventilating pripes as an alternative to carrying them to outside air or to a ventilated discharge stack (ventilated system).

(b) Ventilation stacks serving buildings with not more than ten storeys and containing only dwellings should be at least 32 mm diameter.

(c) The lower end of the stack may be connected directly to a bend or it may be connected to a ventilated discharge stack when the connection should be below the lowest

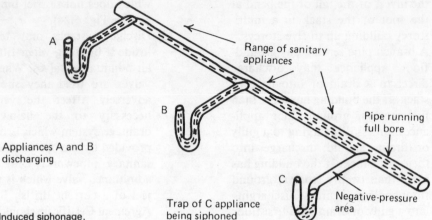

Fig. 6.1 Induced siphonage.

branch discharge pipe (see Fig. 6.8).

(d) The upper end of the stack may be carried to outside air as described (when it should finish as described in 1(b)) or it may be connected to a ventilated discharge stack when the connection should be above the spillover level of the highest appliance.

Principles of Systems

Soil and waste pipe systems should contain the minimum amount of pipework necessary to carry away the foul water from the building quickly and quietly. It should not create a nuisance or a risk to health, nor damage to the building fabric. It must prevent air from the drain or sewer from entering the building under all working conditions.

Loss of Seal in Traps

This can occur by the following means.

Induced Siphonage. This is caused by the discharge of water from another sanitary appliance connected to the same pipe. Water passing the branch pipe connection may draw air out of the branch pipe, creating a partial vacuum and thus causing siphonage of the trap (see Fig. 6.1).

Self-Siphonage. This is caused by a moving plug of water in the waste pipe connected to the trap. As the plug of water moves down the pipe, a partial vacuum is created on the outlet side of the trap, thus causing siphonage of the trap (see Fig. 6.2).

Compression or Back Pressure. As water falls down the stack it carries some air with it

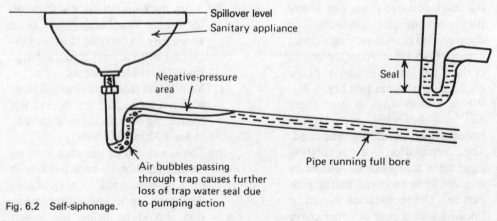

Fig. 6.2 Self-siphonage.

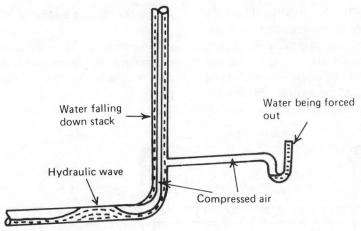

Fig. 6.3 Compression or back-pressure.

and also compresses the air before it. When the water passes through a bend (usually at the base of the stack), the change of direction momentarily slows down the flow of water and a hydraulic wave is also formed in the horizontal pipe. The water flowing behind this hydraulic wave compresses the air and this compressed air may be sufficient to force the seal out of a trap on an appliance close to the bend (see Fig. 6.3).

Capillary Attraction. This is caused by a piece of porous material, such as rag or string, being caught on the outlet of the trap which draws water out of the trap by capillary attraction (see Fig. 6.4).

Wavering Out. If high-velocity wind passes the top of the stack, it may draw some air out of the pipe thus creating a partial vacumm in the pipe. If the wind velocity is variable, the water in the trap oscillates until the seal is broken (see Fig. 6.5).

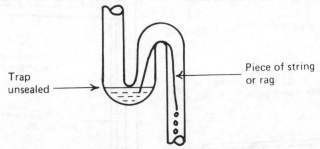

Fig. 6.4 Capillary attraction.

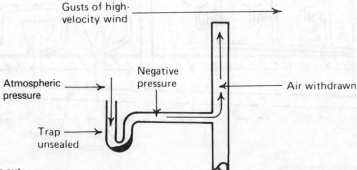

Fig. 6.5 Wavering out.

Evaporation. If the relative humidity inside the building is low and the trap is not used, the seal in the trap may be broken by the water being evaporated. Under normal conditions, the rate of evaporation is about 2.5 mm per week. A trap having a 76 mm seal would lose its seal in about 30 weeks, depend- ing on the relative humidity of the air.

Momentum.   The most usual cause of loss of trap seal by momentum is the sudden dis- charge of a bucket-full of water into a WC pan.

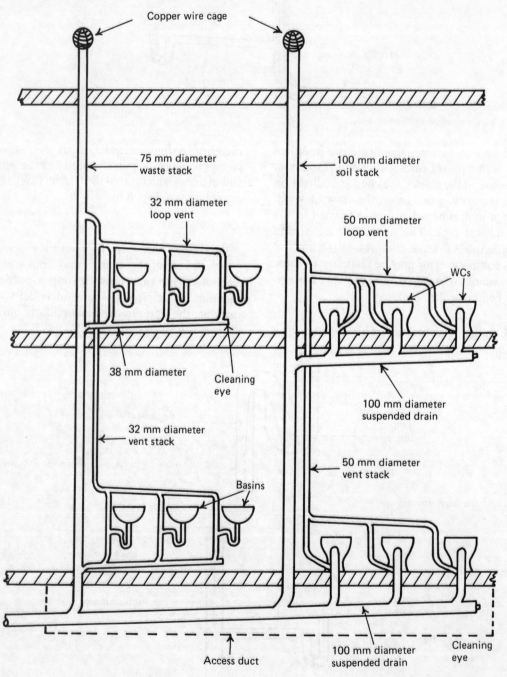

Fig. 6.6   Two-pipe or dual-pipe system.

Leakage. This is usually due to a faulty joint on the cleaning eye or a crack in the trap below the water level.

Systems

Two-Pipe or Dual System (Fig. 6.6). This system is used where there is a long horizontal distance between the sanitary appliances. In buildings such as factories, schools and hospitals, washbasins or sinks may be installed in rooms sited at a long horizointal distance from the main stack carrying the discharges from the WCs. In these cases, it is usually cheaper to install a separate vertical waste stack to the drain for the waste appliances. In the system, waste appliances

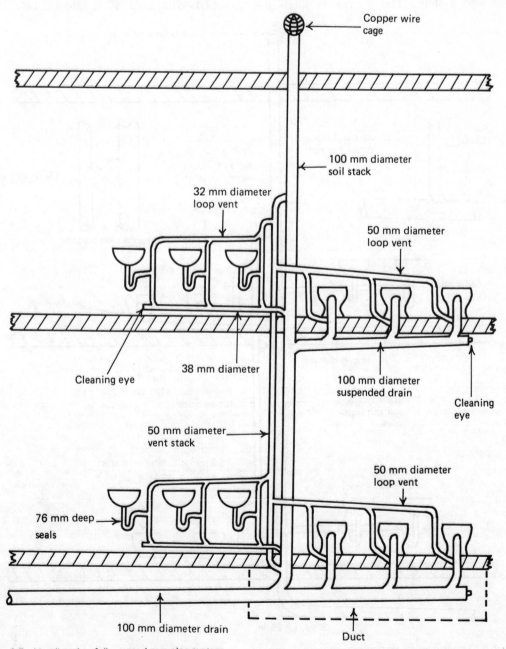

Fig. 6.7 Ventilated or fully-vented one-pipe system.

such as basins, sinks, baths, showers and bidets are connected to the waste stack, and soil appliances such as WCs, urinals, bed-pan washers and slop sinks are connected to the soil stack.

Ventilated System (Fig. 6.7). This system is sometimes referred to as the fully ventilated one-pipe system. The system is used in hospitals, offices and factories where there are a large number of sanitary appliances connected to the horizontal waste or soil branch pipes before connecting them to the vertical discharge stack.

To prevent loss of seals in the traps due to siphonage, an anti-siphon pipe is connected to the outlet of each trap. To prevent back pressure at the base of the stack, the vent stack is connected either into the discharge

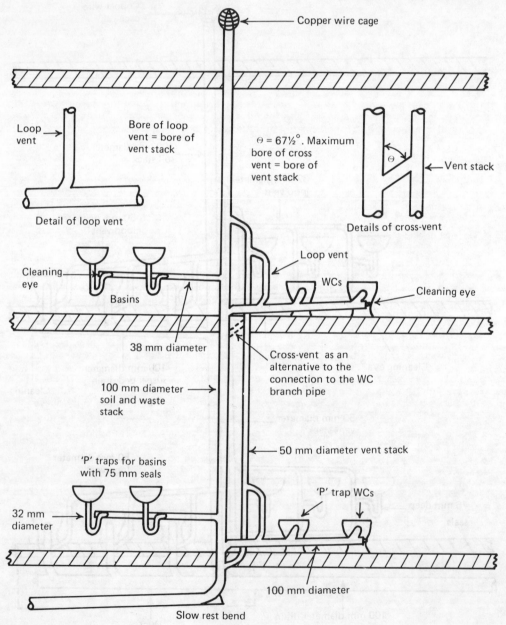

Fig. 6.8 Ventilated-stack system.

stack near to the bend or into the horizontal drain.

Ventilated-Stack System (Fig. 6.8). This system is used in buildings where close grouping of the sanitary appliances makes it practicable, to provide branch discharge pipes without the need for individual venting of traps. Branch pipes, which may serve ranges of up to eight WCs, are normally 100 mm bore and, providing that there are no bends on the branch pipes, anti-siphon pipes are not normally required.

Branch pipes for ranges of urinals are normally 50 or 75 mm bore and therefore anti-siphon pipes are not normally required. A 38 mm bore branch pipe from a bowl urinal, however, should be kept as short as possible.

For ranges of up to four lavatory or washbasins, anti-siphon pipes are not required

providing that the bore of the branch pipe is not less than 50 mm and the slope and length are not greater than 2.5° and 4 m respectively (see Fig. 6.9).

To ventilate the discharge stack, branch ventilating pipes are connected from the soil branch pipes to the ventilating stack. Alternatively, short cross-ventilating pipes may be connected from the discharge stack to the ventilating stack. These branch or cross-ventilating pipes may be connected on each floor or alternate floors, depending upon the discharge stack loading. Table 6.1 gives the recommendations of the Building Research Establishment for ventilated-stack systems.

Modified Single-Stack System (Fig. 6.10). This is sometimes termed the modified one-pipe system, but it is nearer to the design of the single-stack system. It is similar to the ventilated-stack system, but the ventilating

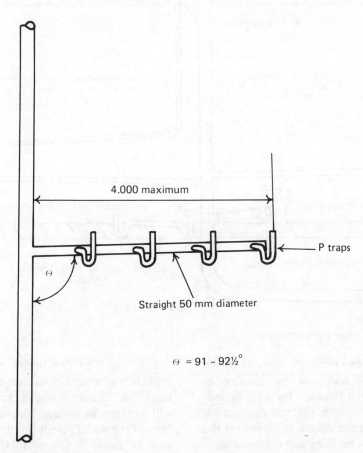

Fig. 6.9 Range of up to four washbasins.

77

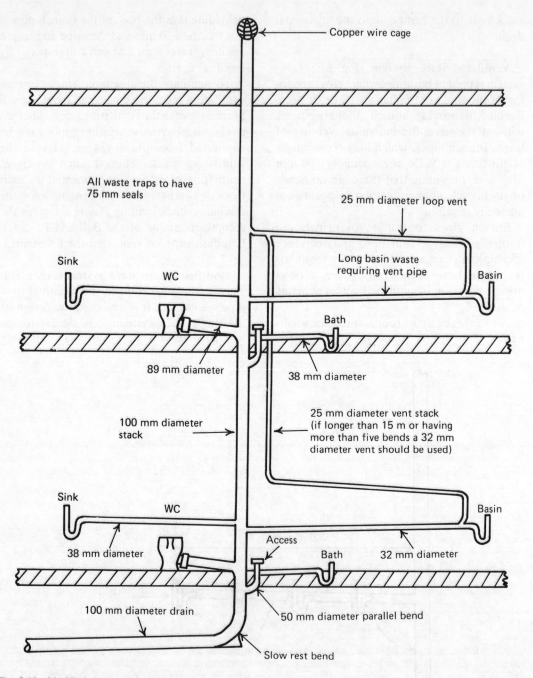

All waste traps to have 75 mm seals

Copper wire cage

25 mm diameter loop vent

Sink

WC

Long basin waste requiring vent pipe

Basin

89 mm diameter

Bath

38 mm diameter

100 mm diameter stack

25 mm diameter vent stack (if longer than 15 m or having more than five bends a 32 mm diameter vent should be used)

Sink

WC

Basin

38 mm diameter

Access

Bath

32 mm diameter

100 mm diameter drain

50 mm diameter parallel bend

Slow rest bend

Fig. 6.10 Modified single-stack system.

stack need not be connected directly to the discharge stack and can be smaller in diameter than that required for a ventilated-stack system. As with the ventilated-stack system, a ventilating stack is provided for the full height of the building and is placed close to the discharge stack.

For ranges of five basins and where the branch waste pipe does not exceed 7 m in length, a 25 mm diameter ventilating pipe will prevent siphonage, providing that the bore of the waste pipe is not less than 50 mm and its slope is not more than 2.5° (see Fig. 6.11).

Table 6.1 Stack Sizes and Cross-Vents Required for the Ventilated-Stack System

Number of storeys	Stack diameter (mm)	Requirements
Flats: stack serving one group on each floor		
6 – 10	100	50 mm vent stack with one cross-vent on alternative floors
11 – 15	100	
12 – 15	125	
16 – 20	100	64 mm vent stack with one cross-vent on each or alternative floors
Maisonettes: stack serving one group on alternate floors		
11 – 15	100	50 mm vent stack with one cross-vent on alternate bathroom floors
16 – 20	100	64 or 50 mm vent stack with one cross-vent on alternate bathroom floors

With spray taps fitted to the basins (without plugs), the discharge rate is only about 0.06 l/s and a 32 mm diameter branch waste pipe may be used. If the number of basins is five, or if the total length of the pipe exceeds 4.5 m, a 25 mm bore ventilating pipe should be provided.

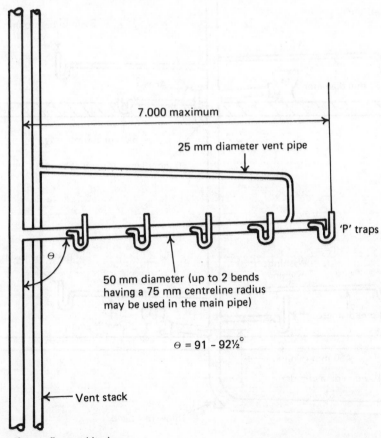

7.000 maximum

25 mm diameter vent pipe

'P' traps

50 mm diameter (up to 2 bends having a 75 mm centreline radius may be used in the main pipe)

$\Theta = 91 - 92\frac{1}{2}°$

Vent stack

Fig. 6.11   Range of up to five washbasins.

Single-Stack System (Fig. 6.12). This system was introduced by the Building Research Establishment and greatly reduces the cost of an installation by designing and installing the pipework so as to eliminate the need for a separate ventilating stack or branch ventilating pipes. The system is usually installed in houses and flats, but may be installed in other types of building.

The main requirements for the system are as follows.

1. Sanitary appliances must be grouped close to the discharge stack so that the branch waste and soil pipes are as short as possible.
2. Sanitary appliances must be individually connected to the discharge stack.
3. The vertical discharge pipe must be

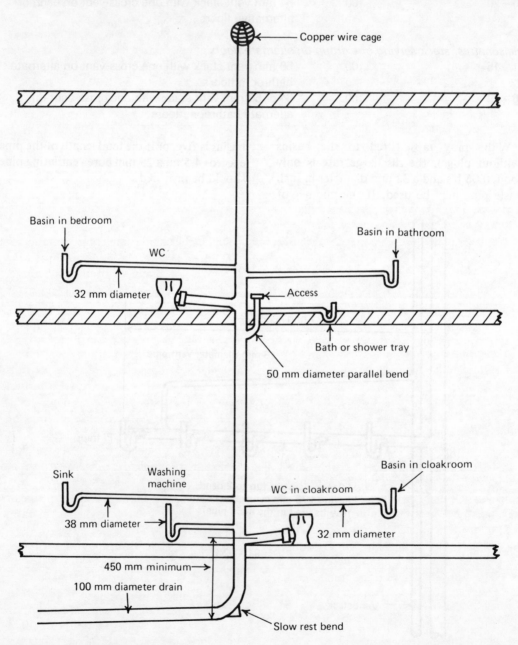

Fig. 6.12   Single-stack system for a house.

80

straight below the highest fitting, to eliminate back pressure in the stack.

4. The bend at the foot of the stack must have a large centreline radius of 200 mm (minimum) or two 135° bends may be used. This is to eliminate back pressure at the base of the stack.

5. For buildings of up to five storeys, the distance from the lowest branch connection to the invert of the drain should be 750 mm minimum. For houses up to three storeys high, this distance should not be less than 450 mm. For large multi-storey buildings, it is usually preferable to connect the ground floor appliances directly into the horizontal drain, instead of into the stack. For buildings above 20 storeys, both the ground and the first-floor appliances may be connected directly into the horizontal drain, instead of into the stack.

6. Junctions for WC connections should be swept in the direction of the flow and the radius at the invert of the junction should not be less than 50 mm.

7. To prevent the discharge from a P-trap WC branch backing up a smaller-diameter bath, or shower branch, the latter should be connected to the stack so that its centreline meets the centreline of the WC branch, or is above this level. Alternatively, the centreline of the bath or shower connection should be at least 200 mm below the centreline of the WC branch connection or an S-trap WC pan may be used (see Fig. 6.13).

8. To prevent the discharge from a small-diameter waste from backing up into another small-diameter waste, the distances between the centre lines of the opposite connections for 75 mm, 100 mm, 125 mm and 150 mm diameter stacks should be 90 mm, 110 mm, 210 mm and 250 mm respectively.

9. If possible, waste appliances should be fitted with P traps.

10. The Building Regulations 1985 state the following maximum lengths of branch waste and soil pipes:
Wash basin; 1.7 m for 32 mm diameter and 3 m for 40 mm diameter.
Sink and bath; 3 m for 40 mm diameter and 4 m for 50 mm diameter.
WC; No limit to length.

11. The Regulations also state the following slopes of branch pipes:
Wash basin; 20–120 mm/m for branch lengths between 600 mm and 1.75 m.
Sink and bath; 18–90 mm/m.
WC; 18 mm/m (minimum).

Note: as the length of pipe increases the slope decreases. This is to prevent the loss of trap water seals by self-siphonage.

Fig. 6.14 shows a single-stack system for a five-storey block of flats.

Stack Diameter

1. 75 or 89 mm diameters are suitable for up to two-storey housing, providing that the diameter of the WC outlet does not exceed the diameter of the stack.

2. 100 mm diameter is suitable for flats of up to five storeys with two groups of fittings on each floor.

3. 125 mm diameter is suitable for flats of up to ten storeys with two groups of fittings on each floor, or of up to 12 storeys with one group of fittings on each floor.

4. 150 mm diameter is suitable for flats of up to 20 storeys with two groups of fittings on each floor.

Note: one group of fittings consist of one or two WCs, with 9 litre flushing cisterns, bath, sink and one or two lavatory basins.

Foaming

The use of detergents produces problems in discharge stacks and even a small amount of detergent can cause a large amount of foam when falling down the stack.

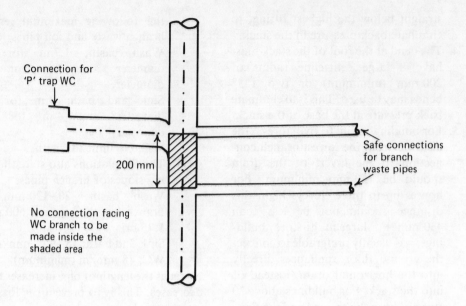

Connection for 'P' trap WC

200 mm

No connection facing
WC branch to be
made inside the
shaded area

Safe connections
for branch
waste pipes

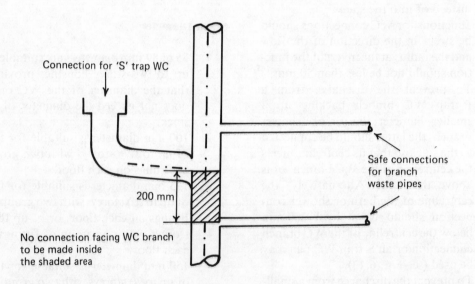

Connection for 'S' trap WC

200 mm

No connection facing WC branch
to be made inside
the shaded area

Safe connections
for branch
waste pipes

Fig. 6.13 Connections near WC branch.

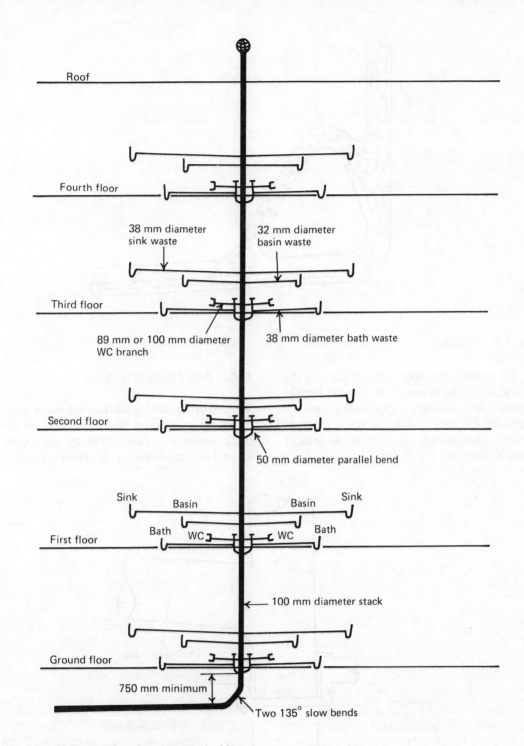

Fig. 6.14 Single stack for a five-storey block of flats.

Labels within figure:

Roof

Fourth floor

38 mm diameter sink waste

32 mm diameter basin waste

Third floor

89 mm or 100 mm diameter WC branch

38 mm diameter bath waste

Second floor

50 mm diameter parallel bend

Sink   Basin   Basin   Sink

First floor   Bath   WC   WC   Bath

100 mm diameter stack

Ground floor

750 mm minimum

Two 135° slow bends

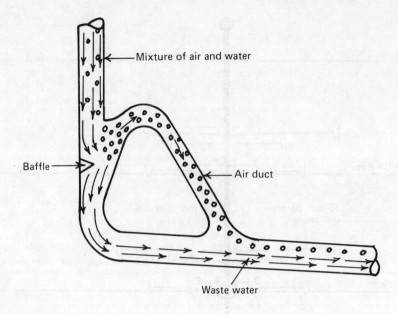

Fig. 6.15   De-aerator.

To prevent excessive back pressure in the stack, it may be necessary to connect certain lower-floor sanitary appliances, via a separate discharge stack, to the horizontal drain. Alternatively, a bottom de-aerator may be used (see Fig. 6.15).

Collar Boss Fitting (Fig. 6.16)

The use of this fitting allows the branch bath or shower tray waste pipe to connect to the stack above the floor level at any point around its circumference. It prevents the dis-

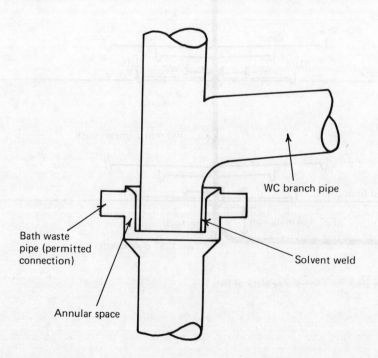

Fig. 6.16   Collar boss fitting.

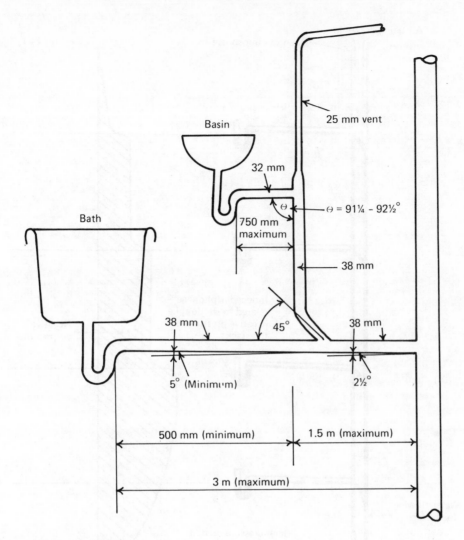

Basin

32 mm

25 mm vent

Θ

$\Theta = 91\frac{1}{4} - 92\frac{1}{2}°$

750 mm maximum

38 mm

Bath

38 mm

45°

38 mm

5° (Minimum)

2½°

500 mm (minimum)

1.5 m (maximum)

3 m (maximum)

Fig. 6.17 Combined bath and basin waste.

charge from the WC backing up into the branch waste pipe.

Combined Branch Pipe (Fig. 6.17)

If required, the bath and lavatory basin may be connected to a common branch waste pipe. However a 25 mm diameter ventilating pipe is required, connected to the basin waste pipe.

Air Pressure in Stacks

In the design of the systems, it is usually acceptable that stacks may be sized to restrict

pressure fluctuations to ± 375 Pa. A negative pressure of this magnitude corresponds roughly to a 25 mm loss of seal from a WC pan, or a 19 mm loss of seal from a small trap with a 75 mm seal. Fig. 6.18 shows the air pressure effects in a discharge stack.

Resealing and Anti-siphon Traps

These are specially designed traps for small unventilated branch pipes fitted to appliances: they prevent loss of trap water seal due to siphonage. They do not, however, prevent loss of trap water seal due to back pressure and will become less efficient if they

85

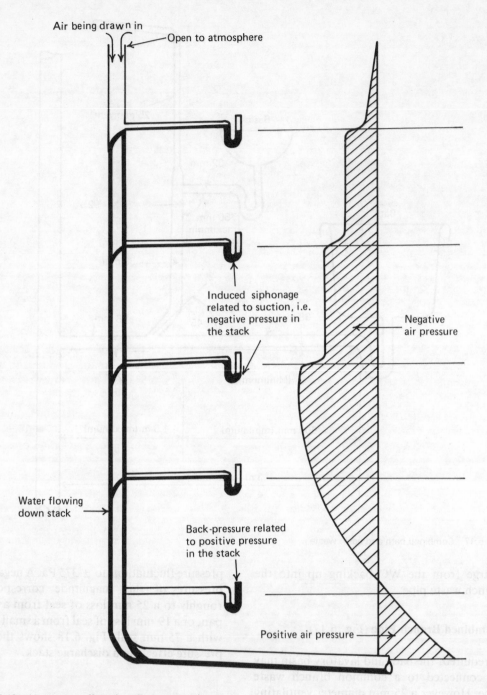

Air being drawn in

Open to atmosphere

Induced siphonage related to suction, i.e. negative pressure in the stack

Negative air pressure

Water flowing down stack

Back-pressure related to positive pressure in the stack

Positive air pressure

Fig. 6.18 Air pressure effects in a discharge stack.

are not regularly maintained. They tend to be noisy and also prevent the thorough ventilation of the branch pipes, which on long lengths may lead to the formation of sediment inside the pipe.

Fig. 6.19 shows a 'Grevak' resealing trap which works as follows.

1. Siphonage of the water in the traps takes place until the water level falls to point A of the anti-siphon pipe.

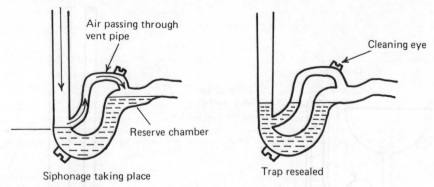

Fig. 6.19 'Grevak' resealing trap.

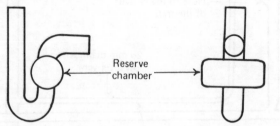

Fig. 6.20 'Econa' resealing trap.

2. Air passes through the vent pipe and equalises the air pressures on the inlet and outlet sides of the trap, thus breaking the siphonic action.

3. Water gravitates from the reserve chamber into the trap and maintains the water seal.

Fig. 6.20 shows an 'Econa' resealing trap which operates as follows.

1. When siphonage of the trap takes place, some of the water is forced by atmospheric pressure into the cylindrical reserve chamber.

2. Water gravitates from the reserve chamber into the trap and maintains the water seal.

Fig. 6.21 shows an anti-siphon trap which is often called an anti-vac trap. The trap works as follows.

1. When water is discharging from either the trap or other traps connected to the same pipe, a negative pressure is created above valve A.

2. The atmospheric pressure acting below the valve forces the valve open and permits air to flow through to the outlet side of the trap, thus preventing siphonic action from taking place.

Low-Level Vent (Fig. 6.22)

It is sometimes necessary to install a sanitary

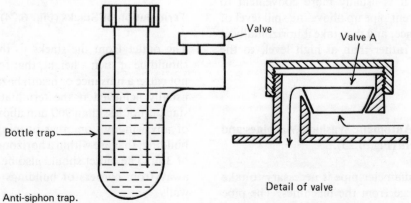

Fig. 6.21 Anti-siphon trap.

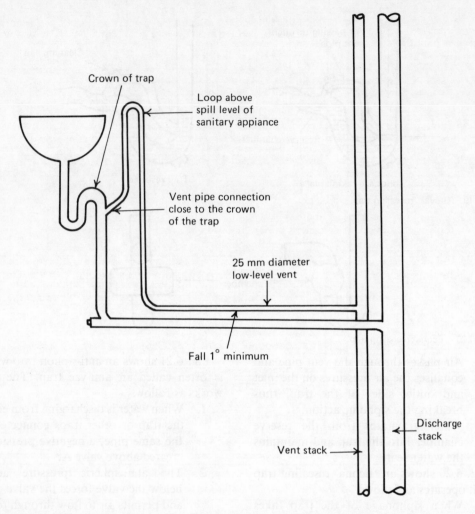

Crown of trap

Loop above
spill level of
sanitary appiance

Vent pipe connection
close to the crown
of the trap

25 mm diameter
low-level vent

Fall 1° minimum

Discharge
stack

Vent stack

Fig. 6.22   Low-level vent (loop vent). Note: if the appliance is on the top floor, the vent pipe may be connected to the discharge stack.

appliance where a room or corridor intervenes between the room in which the appliance is to be installed and the stack. In this case, it is usually more convenient to carry the vent pipe up above the spill level of the appliance, and then take it horizontally at low-level, rather than at high level, to the stack.

Domestic Automatic Washing Machines and Dishwashers (Fig. 6.23).

A 38 mm diameter pipe is necessary to take the discharge from the machines. The pipe

may be connected to the discharge stack or back inlet gully or to a sink branch pipe.

Termination of Stacks (Fig. 6.24)

The outlet from the stacks to the open air should be at such a height, that foul air does not cause a nuisance or health hazard. This is usually achieved if the termination of the stack is not less than 900 mm above the head of any window, or other openings into the building, that are within a horizontal distance of 3 m. The outlet should also be positioned away from corners of buildings or parapet walls.

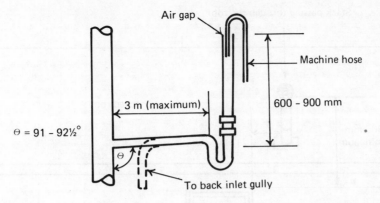

(a) Without use of vent pipe

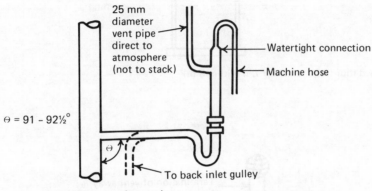

(b) With use of vent pipe

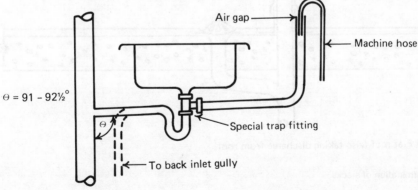

(c) With use of special sink trap fitting

Fig. 6.23 Washing machines and dishwashers. (Reproduced from BS 5572: 1978 by permission of the British Standards Institution, 2 Park Street, London W1A 2BS, from whom complete copies can be obtained.)

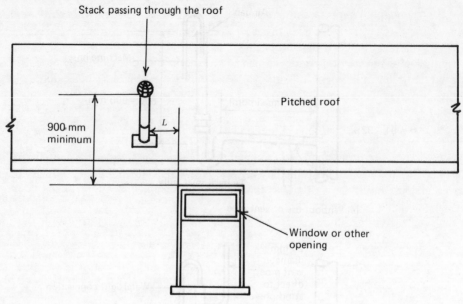

Stack passing through the roof

Pitched roof

900 mm minimum

L

Window or other opening

(a) Pitched roof (requirement if L is within 3 m)

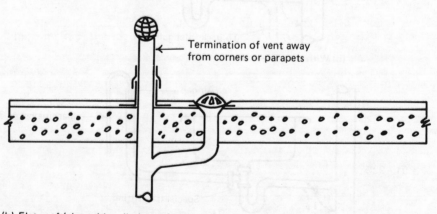

Termination of vent away from corners or parapets

(b) Flat roof (also taking discharge from roof)

Fig. 6.24   Termination of stacks.

Depth of Trap Seals

Traps with outlets for pipes up to and including 50 mm diameter should have a minimum depth of water seal of 75 mm. Traps with outlets for pipes over 50 mm diameter should have a minimum depth of water seal of 50 mm. Traps with a trailing waste discharge and installed on ground

floors, and discharging into an external gully, may have a reduced water seal of not less than 40 mm.

Admission of Rainwater into Stacks

If the discharge from the stack is carried by a combined drainage system, some authorities

90

permit the rainwater from the roofs to be carried by the discharge stack. This method will save a great deal on cost, by eliminating the need for separate rainwater pipes, especially on taller buildings. In very tall buildings, however, for example 30-storey or over, even small continuous flows of rainwater may cause excessive air pressure fluctuations in the discharge stack.

BS 5572:1978, *Sanitary Pipework* recommends that the roof area to be drained by the discharge stacks should be not more than 40 m² per stack, and used for buildings of not more than ten storeys in height.

Pipe Sizing

The usual method of pipe sizing is known as the discharge unit method, in which numerical values are assigned to sanitary appliances to give their load producing properties (see Table 6.2). In this method, the discharge stack is assumed to flow ¼ full (see Fig. 6.25) and branch discharge pipes ½ full.

The discharge unit method cannot, however, be used to find venting requirements, as it is based solely on hydraulic loading and not on air pressure in the stack. A Table is therefore used to give the approximate venting requirements. Table 6.3 gives the maximum capacity and number of discharge units that can be carried by vertical stacks, and Table 6.4 gives the maximum number of discharge units on branch pipes.

Example 6.1. Determine the internal diameter of a discharge stack for a proposed four-storey office, which is to have 5 WCs, 5 basins, 2 urinals and 2 sinks on each floor. With reference to Table 6.2:

```
5 WCs    × 4 ×  14 d.u.'s = 280
5 basins × 4 ×   3 d.u.'s =  60
2 urinals × 4 × 0.3 d.u.'s =   2.4
2 sinks  × 4 ×  14 d.u.'s = 112
                          ─────────
                 Total   = 454.4 d.u.'s
```

With reference to Table 6.3, a 100 mm diameter stack would be suitable; also by

Table 6.2 Discharge Unit Values for Common Appliances

Frequencies given represent the following: 20 min corresponds to peak domestic use; 10 min corresponds to peak commerical use; 5 min corresponds to congested use in public toilets, schools, etc. (Reproduced from BS 5572 : 1978, by permission of the British Standards Institution, 2 Park Street, London W1A 2BS, from whom complete copies can be obtained.)

Type of appliance	Frequency of use (min)	Discharge unit values
WC (9 l flush)	20	7
	10	14
	5	28
Washbasin	20	1
	10	3
	5	6
Spray-tap basin	add 0.06 l/s per tap	—
Bath	75 (domestic)	7
	30 (commercial and congested)	18
Shower	add 0.1 l/s per spray	—
Automtic washing machine	250	4
Sink	20	6
	10	14
	5	27
Urinals (per person)	20 (commercial and congested)	0.3
One group consisting of one WC, one bath, one or two basins, one sink		14

reference to Table 6.4, a 100 mm diameter drain would be suitable having a gradient of 45 mm/m. Alternatively, a 125 mm diameter drain would be suitable having a gradient of 22 mm/m.

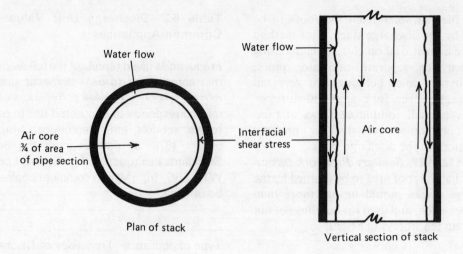

Water flow

Air core
¾ of area
of pipe section

Plan of stack

Water flow

Interfacial
shear stress

Air core

Vertical section of stack

Fig. 6.25 Water flow down stack.

Table 6.3 Maximum Capacity and Number of Discharge Units for Vertical Stacks

Reproduced from BS 5572 : 1978 by permission of the British Standards Institution, 2 Park Street, London W1A 2BS, from whom complete copies can be obtained.

Internal diameter	Approx. capacity of stack (l/s)	Approx. number of discharge units
50	1.2	10
65	2.1	60
75	3.4	200
90	5.3	350
100	7.2	750
125	13.3	2500
150	22.7	5500

Stub Stack (Fig. 6.26)

This may be used for single-storey buildings, providing that the crown of the WC trap is not more than 1.5 m from the invert of the drain and the distance between the highest connection and the invert of the drain does not exceed 2 m.

Table 6.4 Maximum Number of Discharge Units Allowed on Branch Discharge Pipes

Reproduced from BS 5572 : 1978 by permission of the British Standards Institution, 2 Park Street, London W1A 2BS, from whom complete copies can be obtained.

Internal diameter (mm)	Gradient		
	½° (9 mm/m)	1¼° (22 mm/m)	2½° (45 mm/m)
32		1	1
40		2	8
50		10	26
65		35	95
75		100	230
90	120	230	460
100	230	430	1050
125	780	1500	3000
150	2000	3500	7500

Note: Glynwed Foundries Ltd. produce a 200 mm diameter cast iron soil and waste system. In the system, the pipes and fittings are without sockets and are jointed with flexible couplings containing a synthetic rubber gasket. BS 5572 : 1978, *Code of Practice for Sanitary Pipework*, does not include 200 mm diameter stacks, but there is little doubt that the use of this larger-diameter stack will reduce the need for vented stack systems.

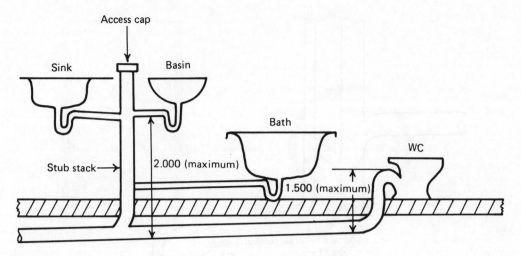

Access cap

Sink

Basin

Bath

WC

Stub stack →

2.000 (maximum)

1.500 (maximum)

Fig. 6.26   Stub stack.

Terminal Velocity in Stacks

In the past, many drainage designers thought that the velocity of flow of water in very high stacks would be excessive and that this would produce excessive noise, damage to pipework and unsealing of traps. In tall buildings, stacks were sometimes offset in order to reduce the velocity of flow; but this was unnecessary, because the force of gravity acting on the water is soon balanced by the resistances set up in the stack.

Terminal velocity in a stack depends on the diameter of the pipe, the volume of flow and the smoothness of the internal bore. The height at which terminal velocity is achieved is very short and is only likely to be equal to one storey. The Building Regulations 1985 state that stacks should not have offsets in any part carrying foul water. If however, an offset is extremely difficult to avoid the regulations may be wavered providing that pressure build-up is prevented in this area. This may be achieved by connecting cross-vents to the discharge stack close to the bends as shown in Fig. 6.27.

Bath Overflow Manifold (Fig. 6.28)

Econa Products Ltd. manufacture a special plastic bath overflow manifold which overcomes the problem of unsightly overflow

pipes from WC flushing cisterns and cold water storage cisterns. The fitting is inserted into the standard overflow hole in the bath. The upper part of the fitting has two 19 mm diameter inlets to which the cistern overflow pipes are connected, as shown in Fig. 6.29.

Testing of Systems

As mentioned earlier in the chapter, the Building Regulations 1985 part H1 requires that a completed soil and waste system should be capable of withstanding a smoke or air test for a minimum period of 3 min, at a pressure equivalent to a head of not less than 38 mm of water. BS 5574 : 1978 gives the following testing procedure.

Air test (see Fig. 6.30). Normally the test should be completed in one operation, but for multi-storey buildings testing in sections may be necessary. The water seals in all the traps should be fully charged and test plugs or bags inserted into the open ends of the pipework to be tested. To ensure that there is a satisfactory air seal at the base of the stack, a small quantity of water to cover the plug or bag can be allowed to enter the system. One of the remaining testing plugs should be fitted with a T piece with a cock on each branch, one branch being connected by means of a flexible tube to a monometer.

Alternatively, a flexible tube from a T piece

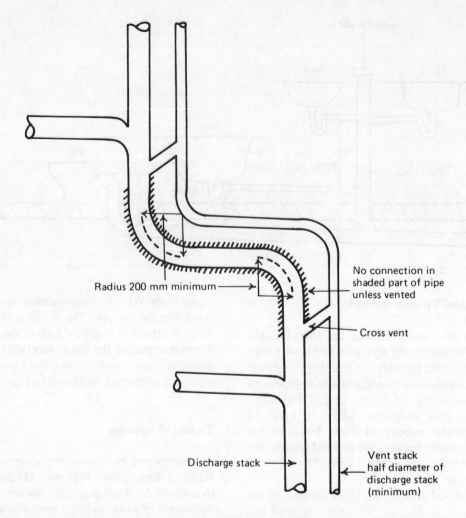

Radius 200 mm minimum

No connection in shaded part of pipe unless vented

Cross vent

Discharge stack

Vent stack half diameter of discharge stack (minimum)

Fig. 6.27 Offset in discharge stack. (Reproduced from BS 5572: 1978 by permission of the British Standards Institution, 2 Park Street, London W1A 2BS, from whom complete copies can be obtained.)

fitted with cocks on its other two branches can be passed through the WC trap. Any water trapped in the tube should be blown out and the manometer connected to one of the branches. Air is pumped into the system until a pressure equal to 38 mm water gauge is obtained. The air inlet cock should then be closed and the pressure in the system maintained for not less than 3 min.

Tests for Maintenance of Trap Seals. In addition to a test for soundness, the system must also be tested to ensure that the trap water seals are also maintained. The Building

Regulations 1985 require that during the test the trap seals should be at least 25 mm.

To test for the effects of self-siphonage of traps, the waste appliances should be filled to overflowing level and discharged in the normal way. To test for individual siphonage and back pressure, a selection of sanitary appliances connected to the stack should be discharged simultaneously. In flats where common discharge stacks are installed, the worst discharge conditions occur when the appliances on the upper floors are discharged. A reasonable test, therefore, is to discharge one WC, one basin and one sink from the top floor of the building, together

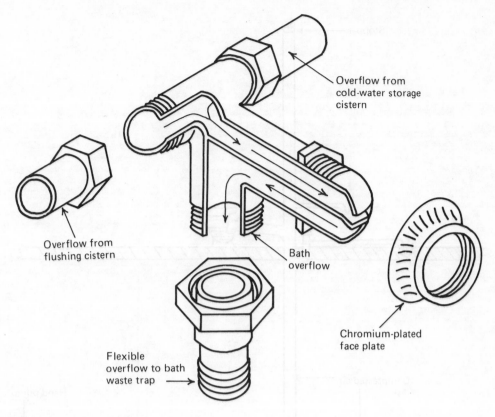

Fig. 6.28 Bath overflow manifold.

Overflow from
cold-water storage
cistern

Overflow from
flushing cistern

Bath
overflow

Chromium-plated
face plate

Flexible
overflow to bath
waste trap

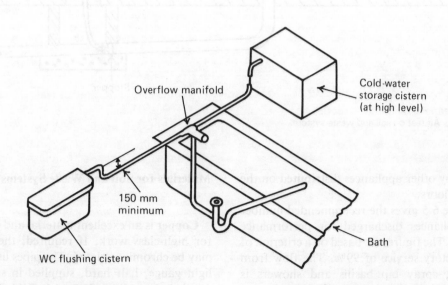

Fig. 6.29 Method of installing bath overflow manifold.

Overflow manifold

Cold-water
storage
cistern
(at high level)

150 mm
minimum

WC flushing cistern

Bath

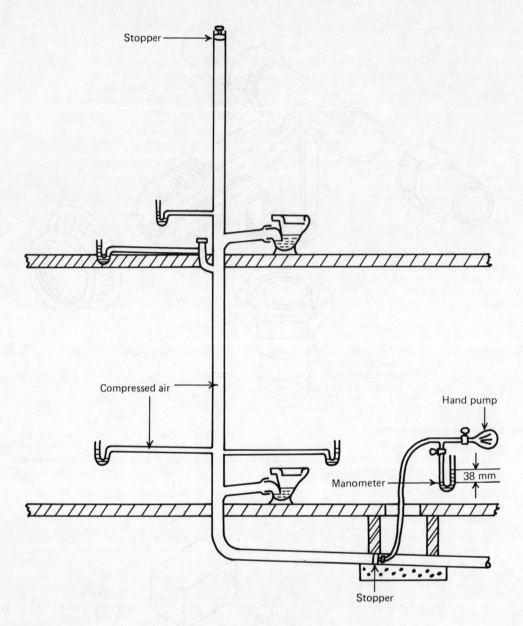

Fig. 6.30   Air test on soil and waste system.

with any other appliances distributed on the lower floors.

Table 6.5 gives the recommended number of appliances discharged for performance testing. The figures are based on a criterion of satisfactory service of 99%. The flow from urinals, spray tap basins and showers is usually small and therefore these appliances need not be discharged.

Materials for Soil and Waste Systems

Copper is an excellent material and is used for high-class work. If required, the pipes may be chromium plated. The pipes used are light gauge, half hard, supplied in straight 6 m lengths to BS 2871 Part 1, Table X. The pipes may be jointed by means of silver

Table 6.5 Number of Appliances to be Discharged for Performance Testing

Reproduced from BS 5527 : 1978 by permission of the British Standards Insitution, 2 Park Street, London W1A 2BS, from whom complete copies can be obtained.

Type of use	Number of appliances of each kind on the stack	Number of appliances discharged simultaneously		
		9 l WC cistern	Washbasin	Kitchen sink
Domestic	1 – 9	1	1	1
	10 – 24	1	1	2
	25 – 35	1	2	3
	36 – 50	2	2	3
	51 – 65	2	2	4
Commercial or public	1 – 9	1	1	
	10 – 18	1	2	
	19 – 26	2	2	
	27 – 52	2	3	
	53 – 78	3	4	
	79 – 100	3	5	
Congested	1 – 4	1	1	
	5 – 9	1	2	
	10 – 13	2	2	
	14 – 26	2	3	
	27 – 39	3	4	
	40 – 50	3	5	
	51 – 55	4	5	
	56 – 70	4	6	
	71 – 78	4	7	
	79 – 90	5	7	
	90 – 100	5	8	

soldering, bronze welding, capillary soldering or compression fittings. (See Fig. 6.31.)

Lead is used for soil and waste branch pipes and has the advantage of adaptability, especially in restricted duct spaces. Lead pipes up to 50 mm in diameter are obtained in coils, and pipes above 50 mm diameter in lengths of up to 3.7 m. The pipes may be jointed by wiped soldered joints or lead welding. (See Fig. 6.32.)

Unplasticised Polyvinyl Chloride (uPVC) produces pipes that are light in weight, easy to handle, smooth in bore and very resistant to corrosion. The pipes expand a great deal when heated and provision should be made to allow for freedom of movement. The pipes

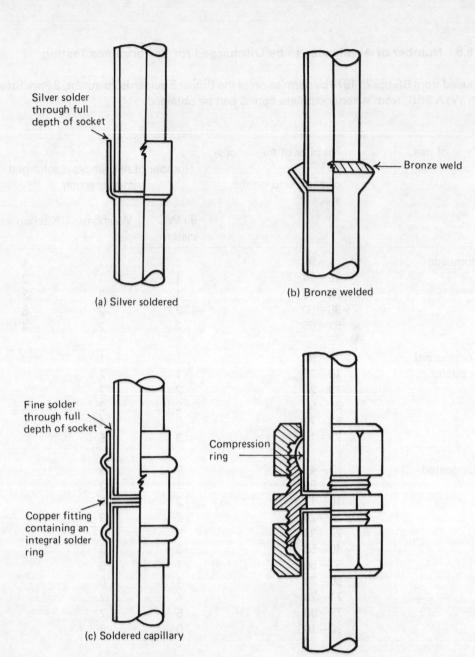

Silver solder
through full
depth of socket

(a) Silver soldered

Bronze weld

(b) Bronze welded

Fine solder
through full
depth of socket

Copper fitting
containing an
integral solder
ring

(c) Soldered capillary

Compression
ring

(d) Compression

Fig. 6.31   Joints on copper pipes.

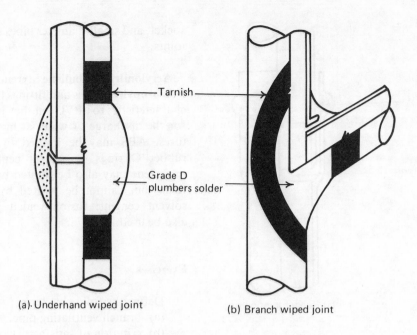

(a) Underhand wiped joint

Tarnish

Grade D
plumbers solder

(b) Branch wiped joint

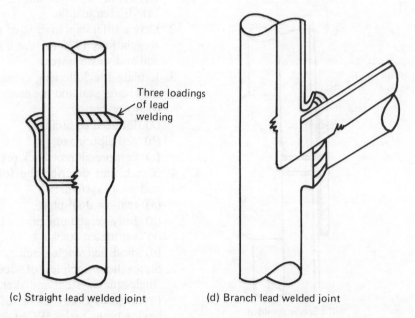

Three loadings
of lead
welding

(c) Straight lead welded joint

(d) Branch lead welded joint

Fig. 6.32   Joints on lead pipes.

are obtained in straight lengths of 10 m and are jointed by means of a rubber ring or solvent cement. (See Fig. 6.33.)

Cast iron will stand up to mechanical damage more than other materials, but it is very heavy and difficult to handle. The pipes

require protection against corrosion by means of a coating of pitch, which is applied to inside and outside surfaces. Pipes may be obtained in lengths of up to 5.5 m and jointed by means of rubber gaskets, rubber rings, caulked lead or caulked lead with a special compound. (See Fig. 6.34.)

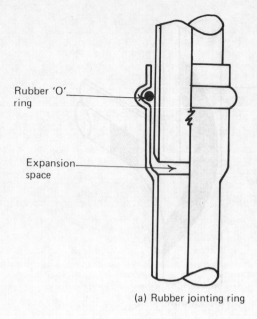

Rubber 'O' ring

Expansion space

(a) Rubber jointing ring

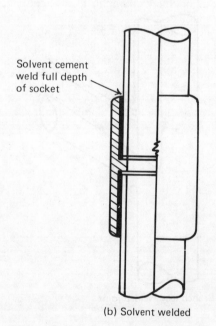

Solvent cement weld full depth of socket

(b) Solvent welded

Fig. 6.33 Unplasticised polyvinyl chloride.

Galvanised Steel is extremely strong and very resistant to mechanical damage. It is much lighter in weight than cast iron and easier to handle. Large-diameter pipes may be jointed by means of a caulked pigot and

socket, and small-diameter pipes by threaded joints.

Acrylonitrile Butadiene Styrene (ABS) and Polypropylene pipes and fittings have similar characteristics to uPVC but they may be used for the discharge of water at high temperatures. ABS may be jointed by means of rubber O rings or solvent cement. Polypropylene may also be jointed by rubber O rings, but cannot be jointed by means of solvent cement; fusion-welded joints may also be used.

Exercises

1. Define the following terms:
    (a) branch ventilating pipe,
    (b) criterion of satisfactory service,
    (c) depth of water seal,
    (d) discharge unit.
2. Give a brief explanation of the Building Regulations 1976 for the installation of soil and waste systems.
3. Explain the following causes of loss of trap water seal and the methods used for their prevention:
    (a) induced siphonage,
    (b) self-siphonage,
    (c) compression or back pressure.
4. Sketch and describe the following soil and waste systems:
    (a) two- or dual-pipe,
    (b) fully-vented one-pipe,
    (c) ventilated-stack,
    (d) modified single-stack.
5. State the principles of design of the single-stack system and sketch a system for a three-storey block of flats each having bath, basin, WC and sink.
6. Sketch the method of installing a range of five washbasins without the use of trap-ventilating pipes.
7. Sketch a diagram to illustrate the air pressure distribution in a discharge stack for a four-storey office having sanitary appliances on each floor connected to the stack.
8. Sketch a section explain the operation of

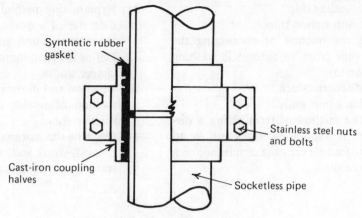

Synthetic rubber gasket

Stainless steel nuts and bolts

Cast-iron coupling halves

Socketless pipe

(a) Synthetic rubber gasket flexible joint

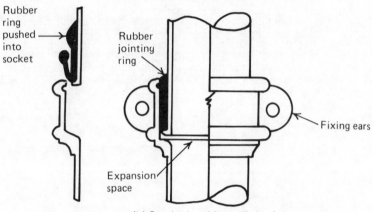

Rubber ring pushed into socket

Rubber jointing ring

Fixing ears

Expansion space

(b) Synthetic rubber roll ring joint

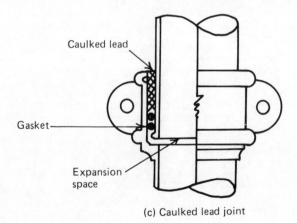

Caulked lead

Gasket

Expansion space

(c) Caulked lead joint

Fig. 6.34 Joints on cast-iron pipes.

(a) a resealing trap,

(b) an anti-siphon trap.

9. Sketch the method of connecting the waste pipe from an automatic washing machine to

(a) a discharge stack,

(b) a back inlet gully.

10. State the method of terminating a discharge or ventilating stack so as to prevent foul air creating a nuisance or a risk to health.

11. Explain the method of calculating the diameter of a discharge stack by use of the discharge unit method.

12. Define the term 'terminal velocity' in discharge stacks.

13. Describe and illustrate with sketches the method of testing completed soil and waste systems.

14. Describe the materials that may be used for soil, waste and ventilating pipes and state the advantages of each.

Chapter 7
Drainage Below Ground, Systems, Ventilation and Access

Principles

An efficient system of drainage to carry away foul and surface water from a building is essential. If the drain is unsound and leaks, the escaping water may be a risk to public health, and may also wash away the soil below the foundations, causing undue settlement of the building.

Drains must satisfy various technical considerations and very often drainage economy influences the siting and planning of a building. Where a public sewer is available, it is usually economical to discharge the drainage into it. If the public sewer passes within 30 m of the site, local authorities may insist that the drainage is connected to it.

Terms Used

The following terms are given by section 343 of the Public Health Act 1936.

DRAIN: a drain used for the drainage of one building, or of any buildings or yards, appurtentant to buildings within the same curtilage.

SEWER: not a drain as defined in this section but, save as aforesaid, includes all sewers and drains used for the drainage of buildings and yard appurtenant to buildings.

PRIVATE SEWER: generally the term drain, sewer and private sewer may be described as follows. A drain is a system of pipes used for the drainage of one or more buildings within a private boundary. The owner of the building or buildings is responsible for the maintenance of the drainage system. A private sewer is a system of pipes used for the drainage of two or more buildings having separate owners. The maintenance of a private sewer is shared by the separate owners.

PUBLIC SEWER: a system of pipes owned and maintained by the local authority. The pipes are usually outside the private boundary and it is the duty of evey local authority to provide public sewers and to provide for the disposal of sewage by means of the sewage disposal plant Fig. 7.1 shows the graphical symbols used for drainage systems.

Systems of Drainage

In some sewage disposal systems, a combined sewer carries the foul and surface water flows. With this method, the sewage disposal plant is often overloaded and additional pumping is sometimes required. In order to reduce the load on the sewage disposal plant, some authorities allow only the foul water flows to be treated, and the surface water is discharged to a water course. The type of drainage system, therefore, will depend on the local authority's sewerage scheme and there are three systems.

Combined. In this system foul water from sanitary appliances and rainwater from roofs and paved areas are conveyed by a single foul

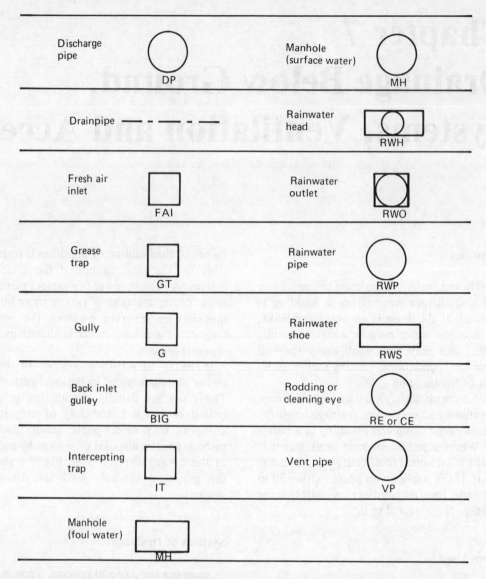

Fig. 7.1 Graphical symbols for drainage, BS 1192.

water drain to a combined sewer. The system saves on drainage cost, but the cost of sewage disposal is increased. Fig. 7.2 shows a combined drainage system for a detached house.

Separate. In this system foul water from sanitary appliances is conveyed by a foul water drain into a foul water sewer and rainwater from roofs and paved areas is conveyed by a surface water drain to a surface water sewer. In rural districts, the local authority may require the rainwater to be discharged

into soakaways. Fig. 7.3 shows a separate drainage system for a detached house.

Partially Separate. This system is basically a separate system, but in order to save on drainage cost an isolated rainwater inlet is allowed to be connected to the foul water drain. In Fig. 7.3, if the rainwater from point A is connected to the foul water drain through manhole B, the system will be changed from a separate system to a partially separate system.

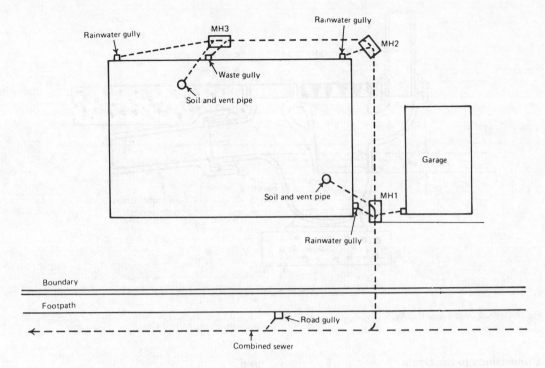

Fig. 7.2 Combined system of drainage.

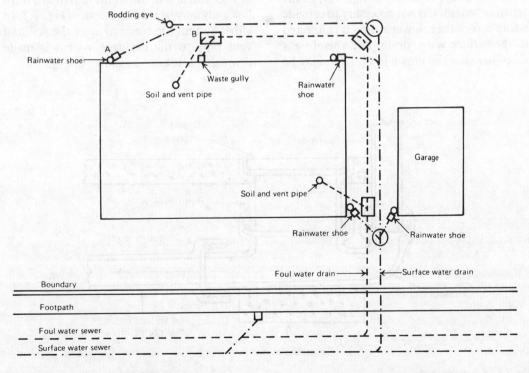

Fig. 7.3 Separate system of drainage.

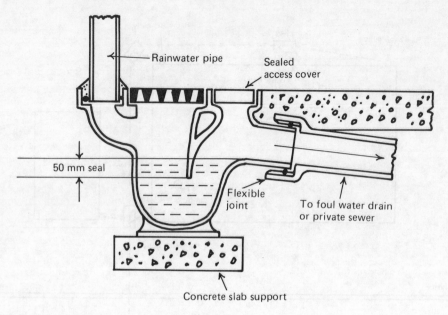

Fig. 7.4   Back inlet trapped gully.

Connections to the Drain

In the combined system, rainwater must be connected to the foul water drain through a back inlet gully, as shown in Fig. 7.4. In the separate system, it is not necessary to provide a trap before the rainwater pipe is connected to the surface water drain, and therefore a rainwater shoe, as shown in Fig. 7.5, may be used.

For the connection of surface water from paved areas into the drains in the combined system, a trapped gully is required as shown in Fig. 7.6, and in the separate system a trapless gully may be used as shown in Fig. 7.7. In either system the connection of the soil and vent pipe to the drainage system is made through a rest bend as shown in Fig. 7.8.

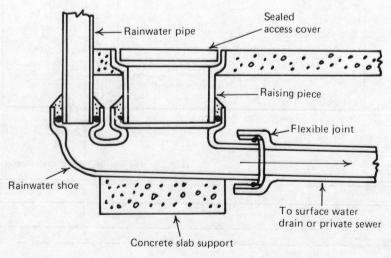

Fig. 7.5   Rainwater shoe.

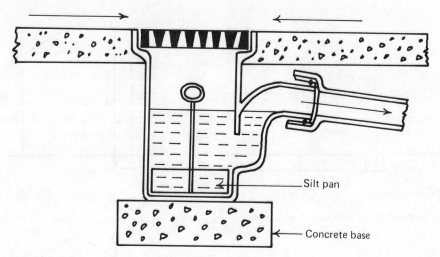

Fig. 7.6   Trapped yard gully.

Silt pan

Concrete base

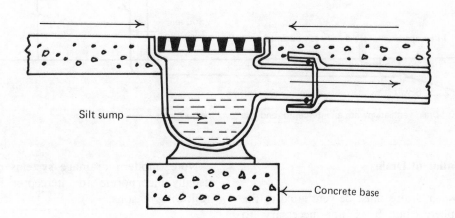

Fig. 7.7   Trapless yard gully.

Silt sump

Concrete base

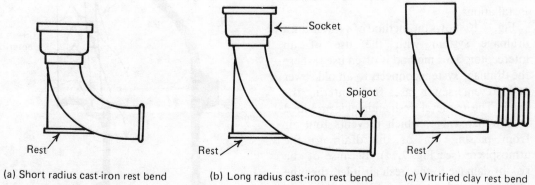

Socket

Spigot

Rest

(a) Short radius cast-iron rest bend

Rest

(b) Long radius cast-iron rest bend

Rest

(c) Vitrified clay rest bend

Fig. 7.8   Rest bends.

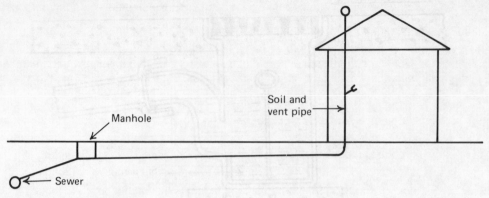

Fig. 7.9   Drain ventilation without an interceptor.

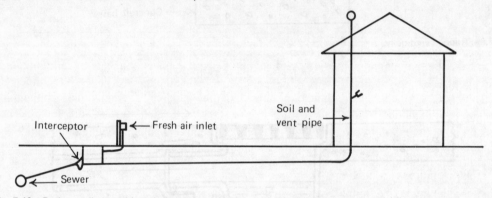

Fig. 7.10   Drain ventilation with an interceptor and low-level fresh air inlet.

Ventilation of Drains

Foul water drains must be ventilated to the atmosphere, but it is not necessary to ventilate surface drains. Fig. 7.9 shows the method of ventilating a drainage system without the use of an interceptor. This method permits the thorough ventilation of the sewer and also reduces the cost of drainage installations.

Fig. 7.10 shows the method of ventilating a drainage system with the use of an interceptor. This method is often used where the drainage system connects to an old sewer and prevents sewer gases from entering the drain. The low-level fresh air inlet is provided with a flap valve, which prevents foul air from passing out of the drain to the atmosphere (see Fig. 7.11). Because of the risk of damage to the fresh air inlet, the pipe may be carried up at high level as shown in Fig. 7.12.

Note: modern drainage systems do not usually incorporate an interceptor for the following reasons.

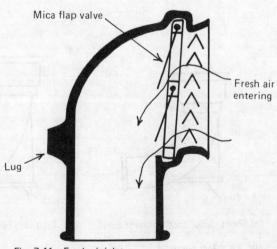

Fig. 7.11   Fresh air inlet.

108

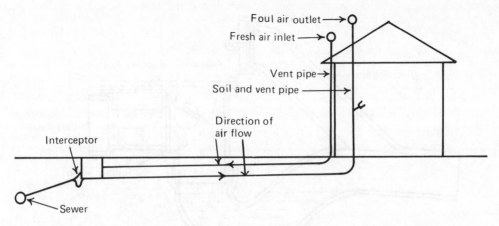

Fig. 7.12 Drain ventilation with an interceptor and high-level fresh air inlet.

1. It prevents the ventilation of the public sewer.
2. It is often the cause of blockage.
3. It increases the cost of drainage.
4. The fresh air inlet, which is usually at low level, is often damaged and the flap valve broken, which then allows drain air to escape to the atmosphere.
5. The rodding arm stopper is sometimes left out and this permits gases from old sewers to enter the drainage system. Fig. 7.13 shows a section through an interceptor.

If there is a risk of the sewage flows backing into the drainage system, an anti-flood interceptor may be used as shown in Fig. 7.14. If back flow from the sewer occurs, the cork float rises and gradually closes the flap valve against the rubber seating, thus preventing the flooding of the drainage system.

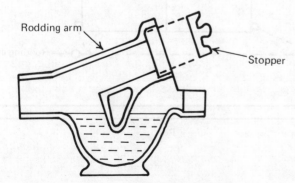

Fig. 7.13 Interceptor.

Use of Private Sewers (Fig. 7.15)

Where several buildings have to be connected to a public sewer, the individual drains may first be connected to a private sewer. This method is cheaper than connecting the individual drains to the public sewer as shown in Fig. 7.16.

Rodding Point System (Fig. 7.17)

The use of this system has the following advantages.

1. It is cheaper than using inspection chambers or manholes.
2. Manholes are a potential source of blockage and sometimes builders' debris enters the drain during their construction.
3. Rodding points are quicker to install than manholes and backfilling and testing can be carried out immediately.
4. A manhole may represent a severe load on the drainage system and may result in a fracture of the drain due to settlement.
5. Vertical rodding points may be formed with lightweight uPVC pipe (see Fig. 7.18).

Manholes

Brick and Precast Concrete. Manholes are

109

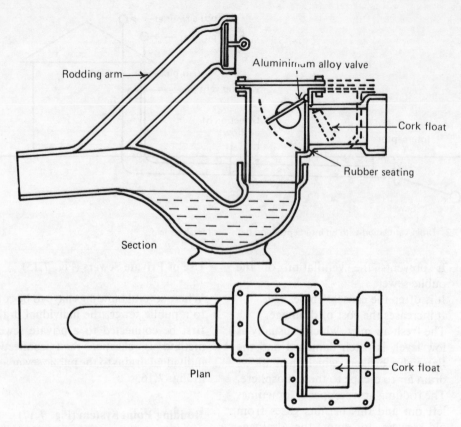

Rodding arm

Aluminium alloy valve

Cork float

Rubber seating

Section

Cork float

Plan

Fig. 7.14   Anti-flood interceptor.

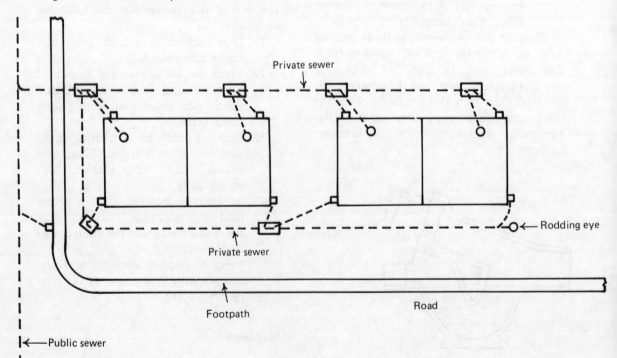

Private sewer

Private sewer

Rodding eye

Footpath

Road

Public sewer

Fig. 7.15   Use of private sewer.

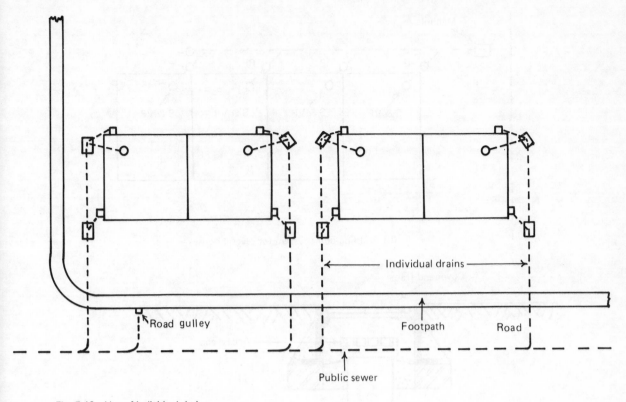

Fig. 7.16 Use of individual drains.

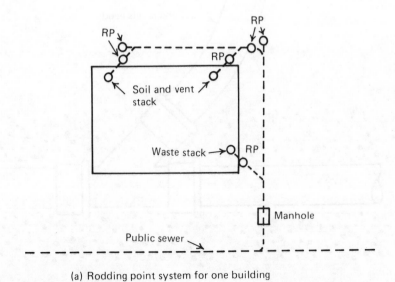

(a) Rodding point system for one building

Fig. 7.17 Rodding-point systems. RP = rodding point.

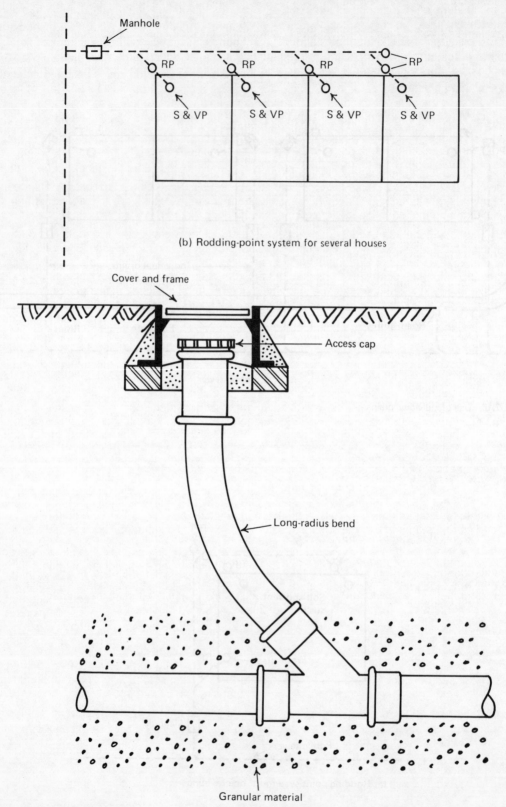

(b) Rodding-point system for several houses

Fig. 7.18 Vertical rodding point.

usually constructed of brickwork, precast concrete or plastic. Brick manholes should preferably the built in English bond, not less than 225 mm thick. Shallow manholes, which are sometimes called inspection chambers, may be built in 113 mm of brickwork, providing that they are not in a road or waterlogged ground. Fig. 7.19 shows a detail

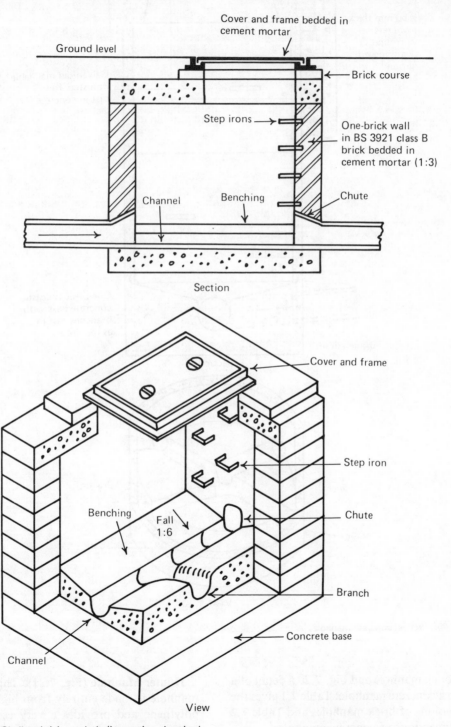

Fig. 7.19   Shallow brick manhole (inspection chamber).

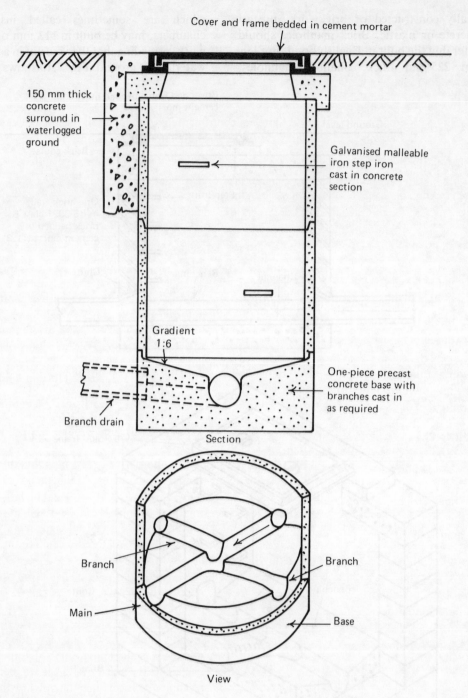

Fig. 7.20 Precast concrete manhole.

Labels in figure:

Cover and frame bedded in cement mortar

150 mm thick concrete surround in waterlogged ground

Galvanised malleable iron step iron cast in concrete section

Gradient 1:6

One-piece precast concrete base with branches cast in as required

Branch drain

Section

Branch

Main

Branch

Base

View

of a brick manhole and Fig. 7.20 a detail of a precast concrete manhole. Table 7.1 gives the dimensions of brick manholes and Table 7.2 the dimensions of concrete manholes.

Hunter Manhole (Fig. 7.21). This type of manhole is made entirely from high-density polythene and provides a very economical and quick method of installation. The man-

Table 7.1 Dimensions of Brick Manholes

Cover sizes for depths up to 2.7 m are 600 mm × 600 mm, and for depths up to 3.3 m are 900 mm × 600 mm. For depths above 3.3 m an access shaft may be constructed above the main chamber.

Type	Depth to invert	Internal length	Internal depth (mm)	Wall thickness (mm)	Concrete base (mm)
Shallow	600 mm	600 mm	450	113	150
	600 mm – 750mm	750 mm	570	225	150
	750 mm – 900 mm	750 mm	700	225	150
	900 mm – 2.7 m	1.2 m	750	225	150
	2.7 m – 3.3 m	1.2 m	750	225	150
Deep	over 3.3 m	1.2 m	750	338	229 – 450

Table 7.2 Precast Concrete Manhole Sections

Rectangular

Internal length (mm)	Internal width (mm)	Wall thickness (mm)	Depth (mm)	Weight (kg)
610	455	51	75	22
610	455	51	100	32
610	455	51	150	45
610	455	51	225	80
610	455	51	300	92
760	610	64	150	70
990	610	64	150	83

Circular

Internal diameter (mm)	Outside diameter (mm)	Approx. weight (kg/m)	Standard lengths (mm)
675	800	310	610, 910, 1220
900	1050	500	610, 910, 1220
1050	1219	650	610, 910, 1220
1200	1397	850	610, 910, 1220
1350	1575	1100	610, 910, 1220
1500	1727	1200	610, 910, 1220
1800	2032	1600	610, 910, 1220
2100	2337	2000	610, 910, 1220
2400	2667	2600	610
2700	3023	3300	910

hole is circular on plan with a diameter of 460 mm. It contains two 110 mm side inlets at 90°, and two side inlets at 45°, and these inlets will accept uPVC pipes or clay pipes with Hepsleve jointing. The base of the chamber contains either 110 mm diameter or 160 mm diameter channels.

Marscar Bowl (Fig. 7.22)

As an alternative to the manhole or rodding point system, an inspection and rodding bowl may be used. The bowl, which is manufactured from uPVC, is fixed at or near ground level and can take up to six 110 mm inlets, which may be connected to the bowl at any desired angle.

The inlets to the bowl have common levels and do not deepen with the fall of the main drain, thus saving the cost of deep excavations. From the bowl, access to the drain line at any depth may be achieved by leading down a 110 mm pipe, and this pipe is connected to the bowl via a 146¼° bend. The pipes are connected to the bowl by means of flanged solvent cement joints. Holes for the connections are made by the use of an Enox cutter.

Backdrops

Where there is a large difference in level

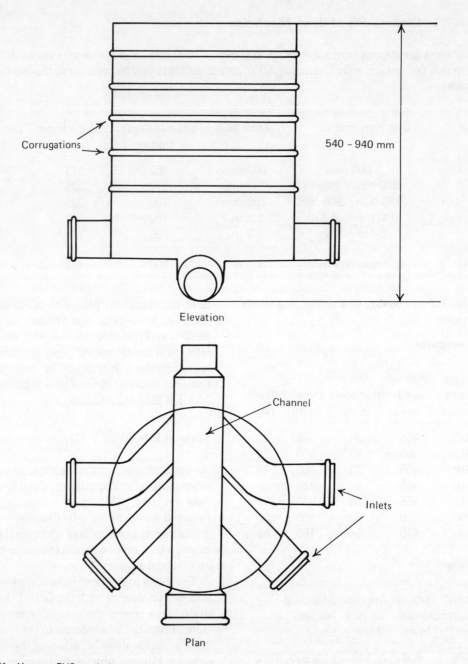

540 – 940 mm

Corrugations

Channel

Inlets

Elevation

Plan

Fig. 7.21 Hunter uPVC manhole.

between the branch drain and the main drain or sewer, a backdrop is required. The backdrop will reduce considerably the amount of excavation required, as shown in Fig. 7.23. In order to save space inside the manhole, branch drains above 150 mm bore have the backdrop usually fixed outside the manhole, using vitrified clay pipe surrounded with concrete 150 mm thick. (Fig. 7.24 shows a backdrop in vitrified clay.) For drains up to 150 mm bore, the backdrop may be fixed inside the manhole, providing there is sufficient space for access. The backdrop inside the manhole may be in pipes made from cast

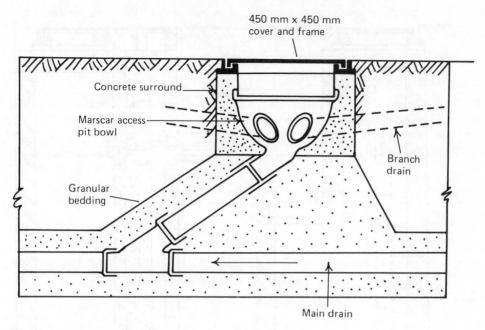

450 mm x 450 mm cover and frame

Concrete surround

Marscar access pit bowl

Granular bedding

Branch drain

Main drain

Fig. 7.22   Marscar bowl.

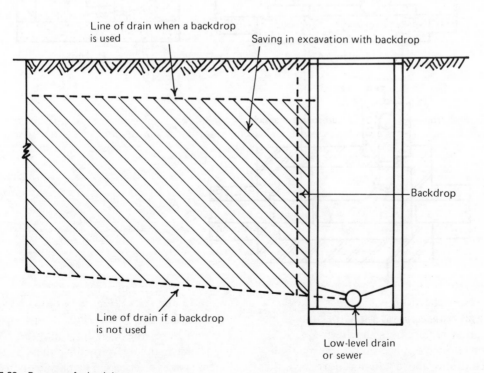

Line of drain when a backdrop is used

Saving in excavation with backdrop

Backdrop

Line of drain if a backdrop is not used

Low-level drain or sewer

Fig. 7.23   Purpose of a backdrop.

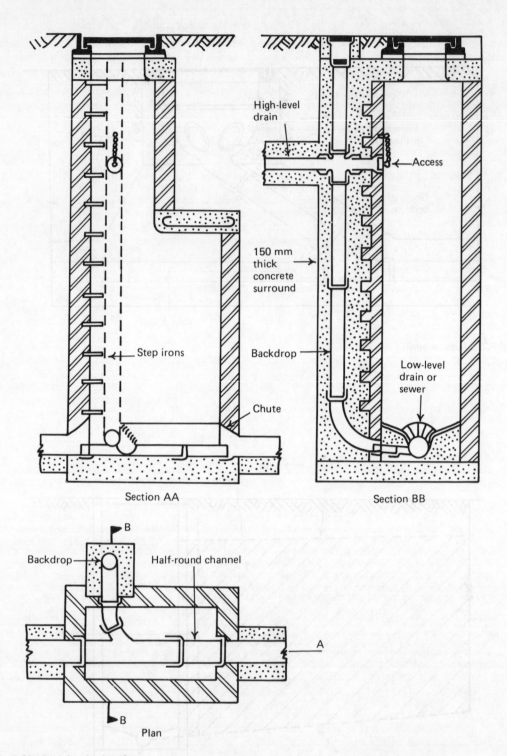

High-level drain

Access

150 mm thick concrete surround

Backdrop

Low-level drain or sewer

Step irons

Chute

Section AA

Section BB

Backdrop

Half-round channel

B

A

B

Plan

Fig. 7.24   Backdrop in clay pipe.

iron, uPVC or pitch fibre. (Fig. 7.25 shows a backdrop in cast iron pipe.)

At one time, backdrops were used to reduce the gradient of branch drains laid below steep sites. In modern drainage practice, however, the drain is laid at the same slope as the steep site, and this method reduces considerably the cost of drainage. It has been found that with steep falls the tendency for solids to be stranded owing to the water running away does not occur, and in fact a steeper fall reduces the risk of blockages. It was also thought that with high flow velocities grit would cause excessive abrasion inside the pipe; with high velocities, however, the turbulent flow carries in suspension the solids which with laminar flow would travel in contact with the surface of the pipe.

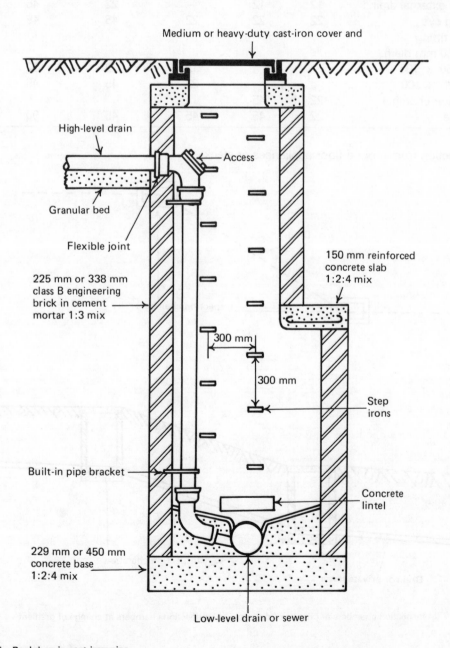

Fig. 7.25 Backdrop in cast-iron pipe.

Table 7.3 Maximum spacing of access points in metres

From	To				
	Access fitting				
	Small	Large	Junction	Inspection chamber	Manhole
Start of external drain*	12	12	—	22	45
Rodding eye	22	22	22	45	45
Access fitting					
small 150 mm diam					
150 × 100			12	22	22
large 225 × 100			22	45	45
Inspection chamber	22	45			
Manhole	22	45	45	45	90

*Connection from ground-floor appliances or stack

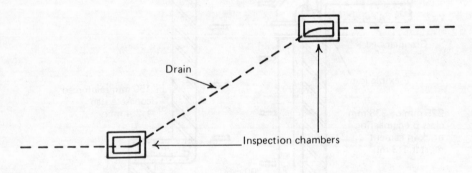

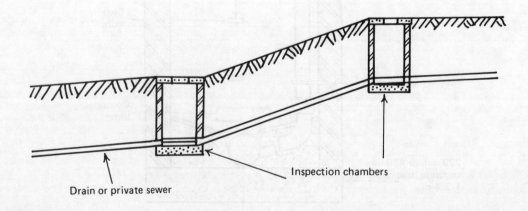

Fig. 7.26   (a) Inspection chambers at change of direction. (b) Inspection chambers at change of gradient.

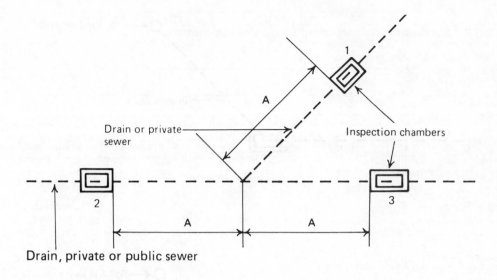

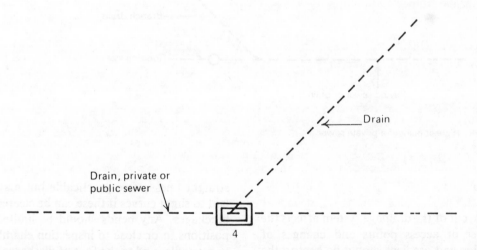

Fig. 7.27 Junctions between drains and sewers. Note: 1,2,3 and 4 are alternative positions of the inspection chambers.

Siting of Access Points

The Building Regulations 1985 require access to drains at the following points:
1. at a bend or change of direction;
2. at a junction, unless each run can be cleared from an access point.

Note: some junctions can only be rodded through from one direction.
3. On or near the head of each drain run;
4. on long runs;
5. at a change of pipe size.

Figs 7.26, 7.27, 7.28 and 7.29 show the positions of access points. The distances marked A depend on the type of access, see Table 7.3.

121

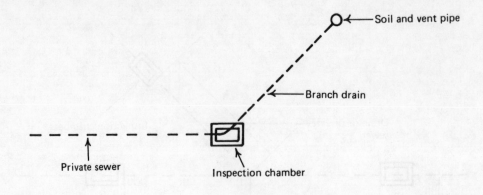

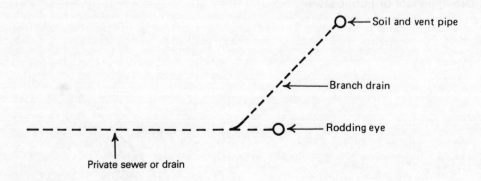

Fig. 7.28   Highest point of a private sewer.

Layout

The layout of the drainage system affects the number of access points and changes of direction and gradient should be kept to the minimum. The pipes should be laid in straight lines where practicable but may be laid to slight curves if these can be cleared of blockages. Any bends should be limited to positions in or close to inspection chambers or manholes and to the foot of discharge and ventilation stacks.

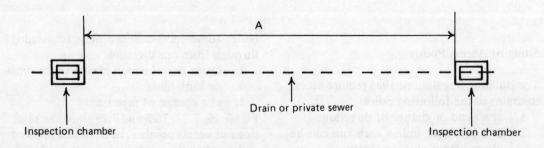

Fig. 7.29   Inspection chambers on a straight run of a drain or private sewer.

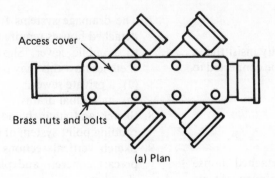

Access cover

Brass nuts and bolts

(a) Plan

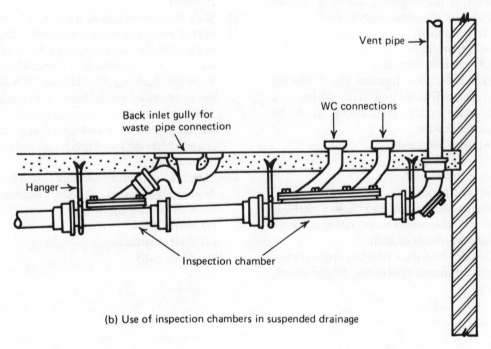

Vent pipe

WC connections

Back inlet gully for
waste pipe connection

Hanger

Inspection chamber

(b) Use of inspection chambers in suspended drainage

Fig. 7.30   Cast-iron inspection chamber.

Pipes should be laid to even gradients and any change of gradient should be combined with an access point.

Special precautions should be taken to accommodate the effects of settlement where drains run under or near a building. Access points should be provided only if blockages could not be cleared without them.

Fig. 7.30a shows the plan of a cast-iron inspection chamber which may be used at the bottom of a manhole or above ground in suspended drainage systems as shown in

Fig. 7.30b. Inspection chambers of this type provide an excellent method of connecting a branch drain to the main drain. The branch drains connect obliquely in the direction of flow and the access cover is bolted down on to a greased felt gasket which provides a water-tight seal. Cast-iron inspection chambers may be obtained with up to six branch connections on each side and bends may be connected to these branches to change the directions of the branch drains.

Exercises

1. When may a local authority insist that a building drainage system be connected to the public sewer?
2. Define the terms
   (a) drain,
   (b) private sewer,
   (c) public sewer.
3. Sketch a plan of a detached house showing the following drainage systems and state the advantages of each:
   (a) combined,
   (b) separate,
   (c) partially separate.
4. Sketch sections through the following drainage fittings and state where in a drainage system each fitting would be used and why:
   (a) back inlet gully,
   (b) rainwater shoe,
   (c) trapless gully,
   (d) rest bend.
5. Sketch and describe two distinct methods of ventilating a drainage system and state the advantages of each.
6. Describe how back flooding from a sewer into a drainage system may be prevented.
7. The drainage systems from three semi-detached houses require to be connected to a public sewer. Show by means of sketches how this may be achieved by
   (a) a private sewer,
   (b) individual drains.
8. Sketch and state the advantages of the rodding point system of drainage.
9. Sketch vertical sections through brick, precast concrete and plastic inspection chambers.
10. State the advantages of a Marscar bowl.
11. Sketch a vertical section through a deep brick manhole incorporating a backdrop and state the purpose of the backdrop.
12. State the Building Regulations 1976 for the position of manholes on a drainage sytem.
13. Sketch a plan of a cast-iron inspection chamber having two branch connections on each side.
14. Sketch a suspended horizontal cast-iron drainage system to take the discharges from the following fittings:
    (a) three WCs,
    (b) four washbasins,
    (c) floor gully.

Chapter 8
Drain Laying, Garage Drainage, Grease Traps

Drain Trench, Levelling

Before commencing the laying of the drains, the bottom of the drain trench must be prepared with the required fall of gradient. The usual method of obtaining the gradient of the trench bottom is as follows.

1. Sight rails are set up, painted black and white, to mark the centreline of the sewer to which the drain is to connect. These sight rails are fixed above the drain trench at the top and bottom of the line of drainage and are given the gradient required for the drain. At least three sight rails should be used, so that if one is accidently disturbed this will be obvious when sighting is

carried out.
2. By means of boning rods, sighted between the sight rails, the level of the trench bottom to give the required gradient may be obtained.
3. Wooden pegs are driven into the trench bottom at 4 m intervals, so that the top of the pegs are level with the bottom of the boning rods.
4. The soil is dug out to about the level of the top of the pegs and a straight edge 4 m long is placed across the pegs, as work proceeds, and the soil levelled to the bottom of the straight edge. After obtaining the gradient required, the pegs are removed for re-use.

Fig. 8.1 shows the method of setting up the

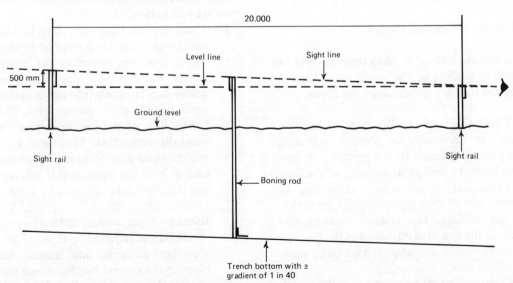

Fig. 8.1   Method of setting up the sight rails. Note: timber supports will be required to support the trench sides.

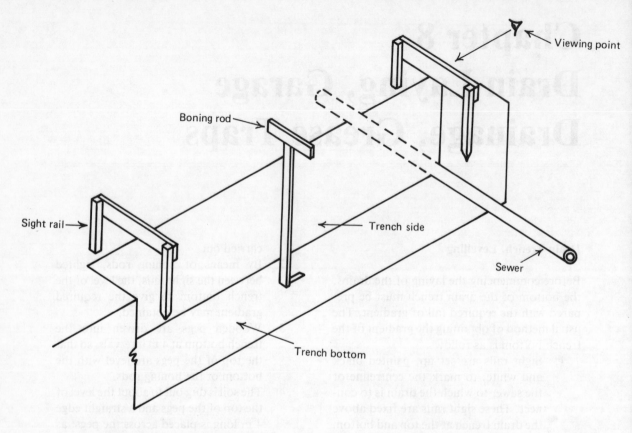

Fig. 8.2 Use of sight rails and boning rods (view).

sight rails and Fig. 8.2 shows the method of using the sight fails and boning rods.

Methods of Bedding Drains

Various methods of bedding the pipes may be used, depending on the condition of the soil and the strength required for the drain.

Class A (Fig. 8.3). This type of bedding may be necessary to provide adequate supporting strength to the pipeline. It may also be used where great accuracy of gradient is demanded, or where it is impractical to remove trench sheeting until after completion of the bedding. This class of bedding also reduces the risk of disturbance of the pipeline if examinations alongside it have to be made at a later date.

Concrete should be spread over the trench bottom to a thickness that will just clear the

pipe sockets; it should be properly compacted and allowed to set.

Two alternative methods of laying the pipes are recommended.

1. Pipes that are light enough to handle easily can be set on a strip of freshly mixed concrete, extending the length of the barrels. The pipes are then gently tapped down into the concrete until they are at the correct level. The consistency of the concrete must be carefully controlled, because if it is too stiff excessive force will be necessary to level the pipes, and if it is too wet the pipes may continue to settle.

2. Heavier pipes may be supported on folding wedges, which can be adjusted to obtain the required level, and concrete poured under and around the pipes to the correct height. When the concrete has set, the wedges should be removed.

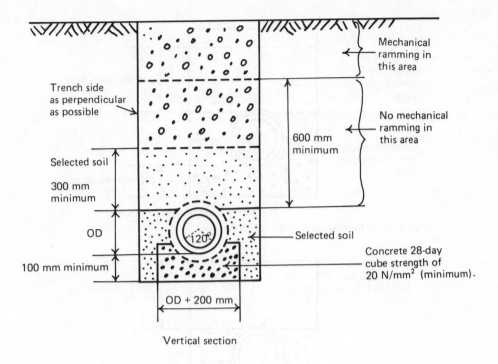

Mechanical ramming in this area

No mechanical ramming in this area

600 mm minimum

Trench side as perpendicular as possible

Selected soil

300 mm minimum

OD

100 mm minimum

Selected soil

Concrete 28-day cube strength of 20 N/mm² (minimum).

120°

OD + 200 mm

Vertical section

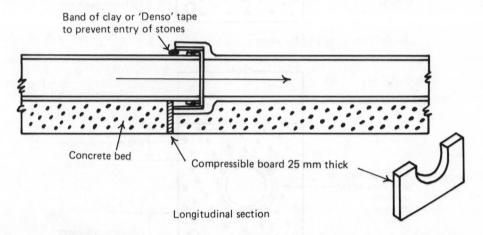

Band of clay or 'Denso' tape to prevent entry of stones

Concrete bed

Compressible board 25 mm thick

Longitudinal section

Fig. 8.3 Class A bedding.

If required, a 150 mm thick layer of concrete may be placed over the pipe to protect it from mechanical damage. Backfilling of the trench should not commence until at least 24 h after completion of concreting and heavy ramming should not be used, nor traffic loads permitted for at least 72 h, or longer in cold weather.

Class B (Fig. 8.4) The use of Class B granular bedding is recommended in preference to laying the pipes directly on the trench bottom. Broken stone or gravel of 10 mm nominal size should be spread and compacted over the trench bottom to a minimum thickness of 100 mm and the top levelled off. The pipes should be laid on this prepared bed, scooping it away to clear the

127

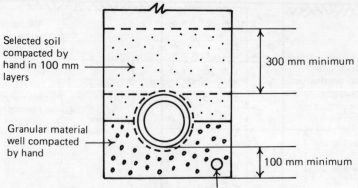

Selected soil compacted by hand in 100 mm layers

300 mm minimum

Granular material well compacted by hand

100 mm minimum

Temporary drain to sump if required

(a) Bedding for clay, cast-iron or concrete pipes

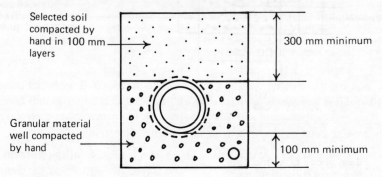

Selected soil compacted by hand in 100 mm layers

300 mm minimum

Granular material well compacted by hand

100 mm minimum

(b) Bedding for pitch fibre pipes

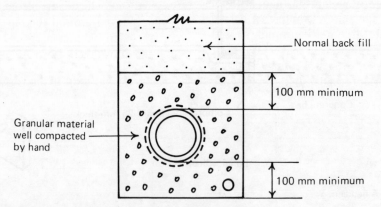

Normal back fill

100 mm minimum

Granular material well compacted by hand

100 mm minimum

(c) Bedding for unplastised polyvinyl chloride pipes

Fig. 8.4 Class B bedding.

usually half-way up the pipe. Any trench sheeting should be removed as work proceeds, so that the granular material fills up the trench faces. After the granular material has been placed, selected soil should be hand-compacted in 100 mm layers up to a level of 300 mm above the crown of the pipes.

sockets. Adjustments to the level can be made by removing or adding bedding material as pipe laying proceeds.

After testing for watertightness, further granular material should be placed equally on both sides of the pipes and compacted in about 100 mm layers up to a specified height,

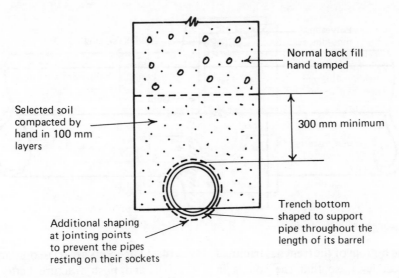

Fig. 8.5   Class C bedding.

Main Backfill. Normally the material excavated from the trench will be used for refilling the trench, but if it is of a type that is difficult to compact and settlement of the surface must be minimised, as for example where a road is to be carried, it may be necessary to use material from another source. It is essential that the material should be placed and spread in layers of not more than 150 mm thickness and adequately compacted before placing the next. Mechanical ramming should not be used until there is at least 600 mm of soil over the pipes. Trench sheeting should be removed as work proceeds, so that spaces are properly filled in.

Note: if required, granular material may be hand-compacted in 100 mm layers up to the level of 300 mm above the crown of the pipes, instead of selected soil.

Class C (Fig. 8.5). In this form of bedding, the soil at the bottom of the trench is shaped to receive a small portion of the pipe barrel. Selected soil, which is free from hard lumps, is then packed around the pipes and hand-compacted in 100 mm layers up to the level of 300 mm above their crown.

Class D (Fig. 8.6). In this form of bedding,

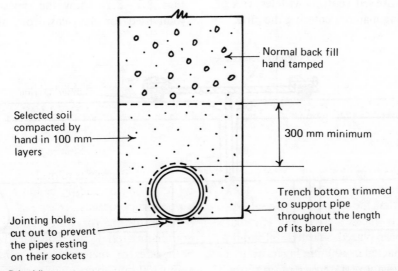

Fig. 8.6   Class D bedding.

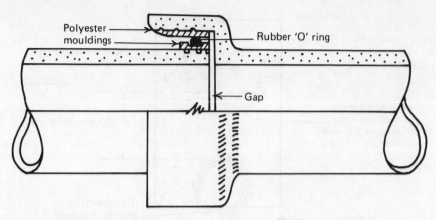

Fig. 8.7 'Hepseal' flexible joint on clay pipe.

the soil at the bottom of the trench is trimmed to the correct level, so that the soil is in contact with the pipe barrel. Socket holes are dug to the correct depth, so that the pipes rest on their barrels and not on their sockets. Selected soil is then packed around and above the pipes as described for Class C bedding.

Jointing of Pipes

Flexible joints on drainage pipe lines are strongly recommended in preference to caulked lead, cement mortar or other forms of rigid joint. They have the following advantages.

1. They are quicker and easier to joint.
2. They are very reliable and cheap.
3. They are self-centring with less risk of jointing material entering the pipe.

4. The joints are telescopic and resist pull and push fracture better than rigid joints; also better with regard to expansion and contraction.
5. A test may be applied immediately after the pipes have been laid.
6. They permit movement of the ground, due to shrinking of clay or settlement, without fracturing.
7. Because they are quicker to make, the time that the trench is left open is reduced to a minimum, with a possible saving in pumping and also less risk of the trench bottom becoming water-logged or of damage to the pipes by vandals.
8. There is less interruption of laying during wet or freezing weather.

Figs. 8.7 – 8.14 show the types of flexible joint used for clay, cast-iron, uPVC, pitch

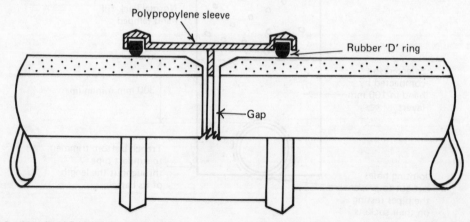

Fig. 8.8 'Hepsleve' flexible joint on clay pipe.

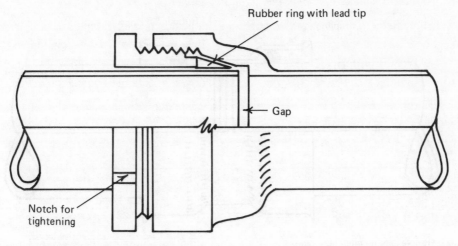

Fig. 8.9   Screwed-gland flexible joint on cast-iron pipe.

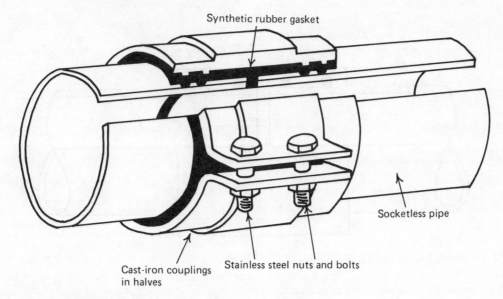

Fig. 8.10   'Glynwed' flexible joint on cast-iron pipe.

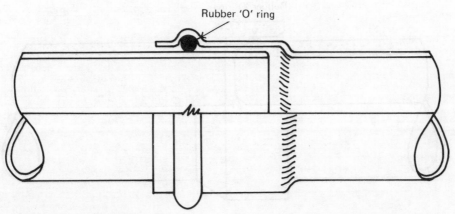

Fig. 8.11   Flexible joint on uPVC pipe.

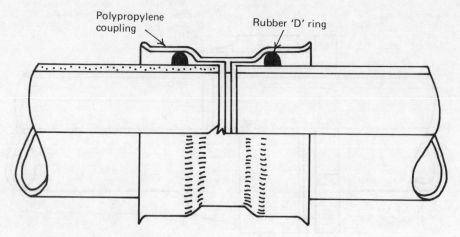

Fig. 8.12   Flexible joint on pitch fibre pipe.

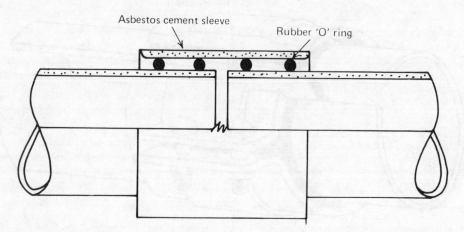

Fig. 8.13   Flexible joint on asbestos cement pipe.

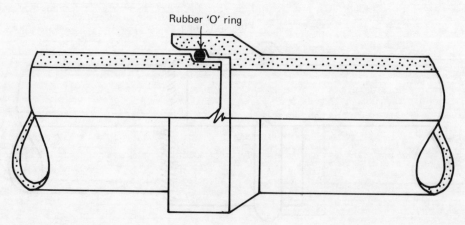

Fig. 8.14   Flexible joint on concrete pipe.

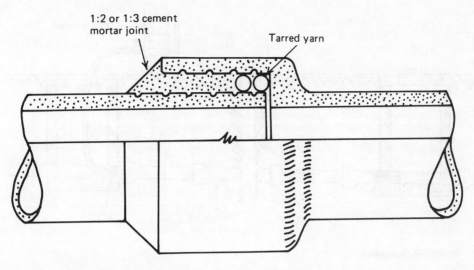

1:2 or 1:3 cement mortar joint

Tarred yarn

Fig. 8.15 Rigid cement mortar joint on clay pipe.

fibre, asbestos cement and concrete pipes. Figs. 8.15 and 8.16 show rigid joints on clay and cast-iron drain pipes.

Building Regulations and Code of Practice Tests

The Building Regulations 1985 and the Code of Practice BS 8301, Building Drainage 1985 give the following procedures for carrying out tests on gravity drains and private sewers up to 300 mm in diameter.

Water test.   The drain should be fitted with water to give a test pressure equal to 1.5 m head of water above the soffit of the drain at the high end but not more than 4 m head of water above the soffit of the drain at the low end. Steeply graded drains should be tested in stages so that the head of water at the lower end does not exceed 4 m. This is to prevent damage to the drain and it may be necessary to test a drain in several sections.

The pipeline should be allowed to stand for 2 hours for absorption and topped up with water as necessary. After 2 hours the loss of water from the pipeline should be measured by noting the quantity of water needed to

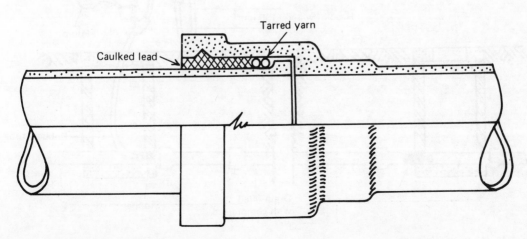

Tarred yarn

Caulked lead

Fig. 8.16 Rigid caulked lead joint on cast-iron pipe.

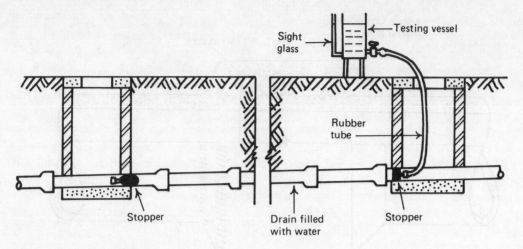

Fig. 8.17 Water test on drainage.

maintain the test head for 30 minutes. The reate of water loss should not exceed 1 litre/hour per metre diameter, per metre run of pipe. For various pipe diameters the rate of loss per metre run during the 30-minute period is 0.05 litre for 100 mm pipe; 0.08 litre for 150 mm pipe; 0.12 litre for 225 mm pipe and 0.15 litre for 300 mm pipe. Fig. 8.17 shows the method of carrying out the water test using a rubber tube connected to a testing vessel.

Alternatively, the test may be carried out by means of a temporary bend and standpipe connected to the head of the drain. The drain

should be strutted to prevent movement of the drain during testing.

The fall of the water level in the testing vessel or standpipe may be due to
1. absorption by pipes and joints,
2. sweating of pipes and joints,
3. leakage from defective pipes or joints,
4. leakage from testing stoppers,
5. trapped air.

Some pipes absorb more water or trap more air at the joints than others and an allowance is therefore made for this by adding water to maintain the test head during the period of 30 minutes.

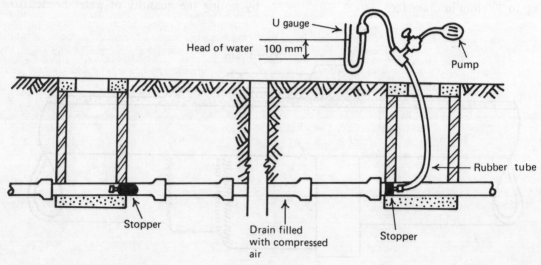

Fig. 8.18 Air test on drainage.

Note: if there is a leakage from testing stoppers, the stopper should be tightened. If an inflatable bag stopper is used, more air should be pumped into the bag.

Sometimes a leak can occur on the threaded portion of the plug. All equipment used for testing must be thoroughly checked before being used and rubber surfaces of stoppers should be free from grit.

Air test. The length of drain to be tested should be plugged and air pumped into the pipe until a pressure of slightly more than 100 mm water gauge is obtained. Where gullies and/or ground-floor appliances are connected, the test pressure should be slightly more than 50 mm water gauge. A change in air temperature will affect the test pressure and therefore 5 minutes should be allowed for pressure stabilisation. The air pressure should be adjusted to 100 mm or 50 mm water gauge as necessary. Without further pumping, the head of water during a period of 5 minutes should not fall more than 25 mm and 12 mm for 100 mm and 50 mm water gauge test pressures respectively.

Figure 8.18 shows the method of carrying out an air test. The test is carried out by fixing a stopper, sealed with a cap, at one end of the drain and pumping in air at the other end until the U gauge shows the required head of water.

Smoke Test (Figs. 8.19 and 8.20). This test is useful for drainage both above and below

ground and a leak on an exposed pipe can be easily seen by the escape of smoke. If the drain is covered, it is sometimes possible that smoke might percolate through the soil to indicate the position of a leak.

The test is prepared by effectively plugging the length of drain to be tested and connecting a smoke machine to the lowest stopper. The test is carried out by forcing smoke into the drain, by means of a smoke machine, until the dome rises to the height required. During the test, the dome should remain stationary and the drain inspected for any trace of smoke. Figs. 8.21 and 8.22 show the types of stopper used for the tests.

Test for Straightness and Obstruction (Fig. 8.23). This test can be carried out by placing a mirror at one end of the drain and a lamp at the other end. It will be seen by looking through the mirror whether or not the drain is straight or has an obstruction.

Manholes. If the ground water level is likely to be seasonally above the soffit of the drain, manholes should be inspected for watertightness against infiltration when the water table is at its highest. Petrol interceptors should be tested for watertightness by filling with water before backfilling is commenced. After a reasonable period has elapsed, the water level should be made good by adding further water as required. This water level should be maintained for 30 min without adding further water.

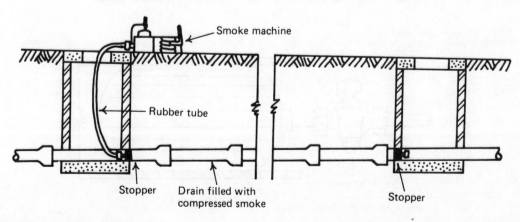

Fig. 8.19 Smoke test on drainage.

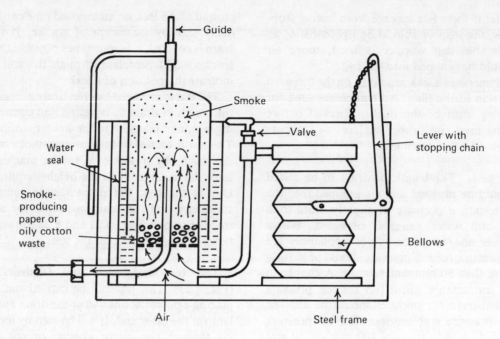

Fig. 8.20   Smoke machine during testing.

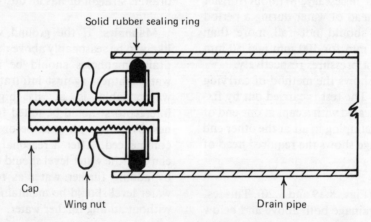

Fig. 8.21   Expanding rubber drain stopper. Note: cap can be removed and hosepipe connected.

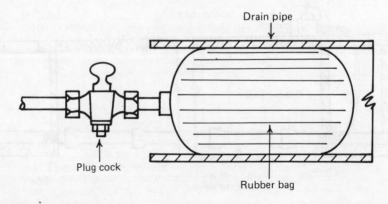

Fig. 8.22   Inflatable bag drain stopper.

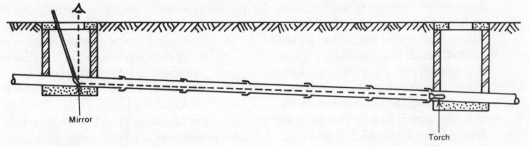

Fig. 8.23 Test for straightness and obstruction.

Trenches For Drains and Private Sewers
(Fig. 8.24)

The Building Regulations 1985 require that:

1. Where any drain or private sewer is constructed adjacent to a load-bearing part of a building, such precautions shall be taken as may be necessary to ensure that the trench in which the

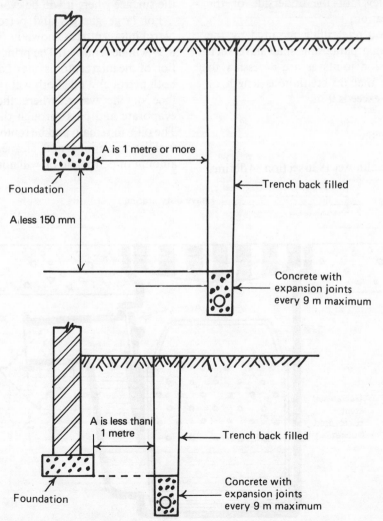

Fig. 8.24 Trenches for drains and private sewers.

drain or private sewer is laid in no way impairs the stability of the building.

2. Except that where the nature of the ground makes it unnecessary, where any drain or private sewer is adjacent to a wall and the bottom of the trench is lower than the foundation of the wall, the trench shall be filled in with concrete to a level which is not lower than the bottom of the foundation of the wall by more than the distance from that foundation to the near side of the trench less 150 mm.

But, where the trench is within 1 m of the foundation of the wall, the trench shall be filled in with concrete to a level of the underside of the foundation.

3. The concrete filling required by the foregoing paragraph shall have such expansion joints as are necessary to ensure that no continuous length of filling exceeds 9 m.

Garage Drainage

The public Health Act 1936 section 34 defines certain prohibited discharges into drains or sewers as

1. anything that may injure a drain or sewer or interfere with the free flow or treatment and disposal processes,
2. hot liquids with a temperature exceeding 43.3 °C,
3. petroleum spirit and calcium carbide.

This means that the floor washings of large garages, petrol stations and indeed small garages should be provided with some means of intercepting petrol before it enters the drain or sewer. For the floor washings of a small garage, it is sufficient to provide a garage gully as shown in Fig. 8.25. This will retain mud and oil, and petrol will float on the surface where it will be evaporated.

For large garages and petrol stations, a petrol interceptor as shown in Figs. 8.26 and 8.27 will be required. The principle of operation of the interceptor is one of elimination of both petrol and oil. Both will rise to the surface of the water, where the petrol will evaporate and pass through the vent pipes. The oil will remain and be removed when the tanks are emptied and cleaned. A certain amount of petrol and oil will not have time to

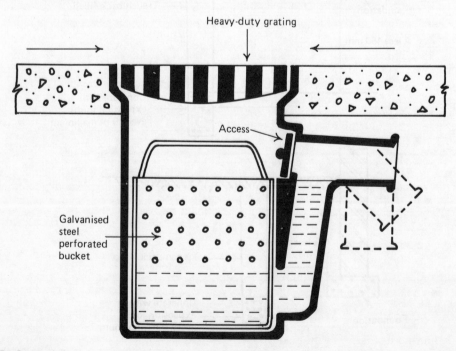

Fig. 8.25 Garage gully.

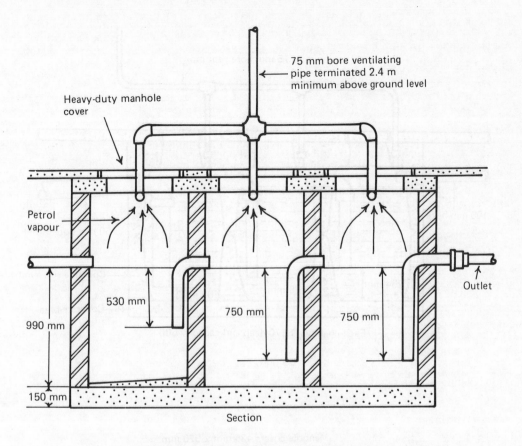

75 mm bore ventilating pipe terminated 2.4 m minimum above ground level

Heavy-duty manhole cover

Petrol vapour

990 mm

530 mm

750 mm

750 mm

Outlet

150 mm

Section

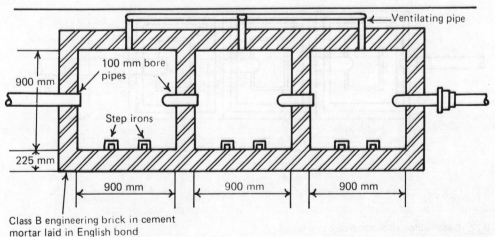

Ventilating pipe

900 mm

100 mm bore pipes

Step irons

225 mm

900 mm

900 mm

900 mm

Class B engineering brick in cement mortar laid in English bond

Fig. 8.26 Brick petrol interceptor.

separate out in the first chamber and an additional one or two chambers are therefore required. Fig. 8.28 shows the plan of the floor drainage for a medium-sized garage. The area of the surface to be drained by each gully should not usually exceed 50 m².

Grease Traps (Fig. 8.29)

Special gullies for the collection of grease are not required for houses, but for canteen kitchens where the waste water from the sinks and dishwashers contains a considerable

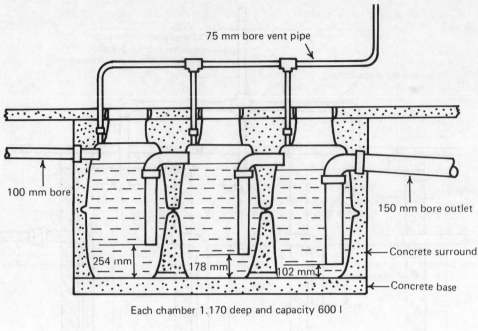

75 mm bore vent pipe

100 mm bore

150 mm bore outlet

Concrete surround

Concrete base

254 mm

178 mm

102 mm

Each chamber 1.170 deep and capacity 600 l

Section

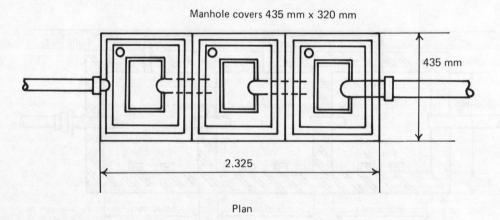

Manhole covers 435 mm x 320 mm

435 mm

2.325

Plan

Fig. 8.27   Glass-reinforced plastic petrol interceptor.

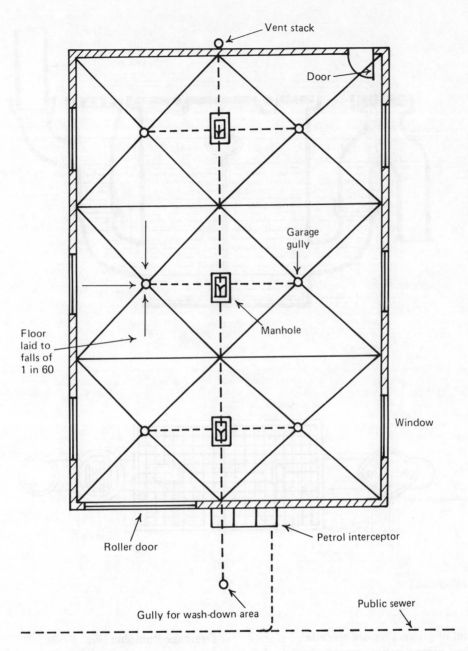

Fig. 8.28 Plan of garage drainage.

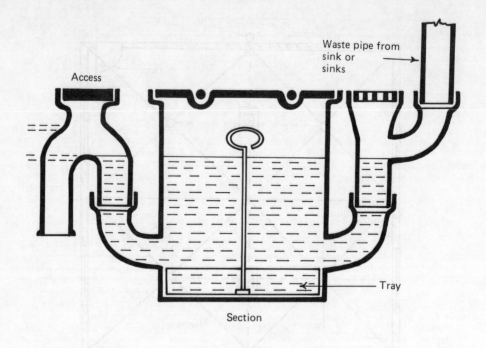

Access

Waste pipe from sink or sinks

Tray

Section

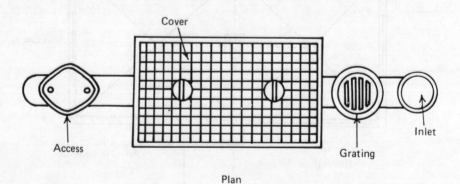

Cover

Access

Grating

Inlet

Plan

Fig. 8.29   Grease trap.

amount of grease they are essential.

When grease is hot or contained in hot water, it is in the form of an emulsion, and if it is allowed to flow into the drain it will cool and adhere to the sides of the pipes. The principle of operation of the grease trap is that of cooling down the grease in a large volume of water, which will generally be cool, so that the grease is solidified and floats on the surface. At periodic intervals, the tray is lifted out of the trap, which at the same time collects the grease.

Drainage Pumping (Fig. 8.30)

Wherever possible, drains should be laid so that the liquid flows by gravity to the sewer, or other point of disposal. In some cases, however, the sewer or point of disposal is above the drain, and pumping is therefore required. For the pumping of both foul water and surface water, a pumping installation as shown in Fig. 8.30a may be used. For larger installations, two pumps should be installed, so that one of the pumps may be used for

142

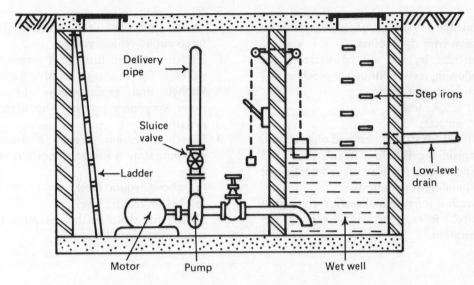

(a) Pump installation for foul and surface water

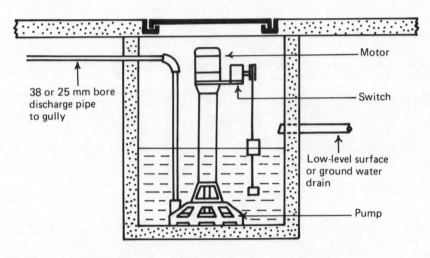

(b) Sump pump for surface or ground water

Fig. 8.30 Drainage pumping.

stand-by purposes.

For the pumping of surface or ground water, a sump pump installation may be used as shown in Fig. 8.30b. This type of installation is used for basements and boiler rooms to deal with seepage of water, floor washing or the draining down of the boilers and heating pipework.

Exercises

1. Sketch and describe the method of obtaining the gradient of a drain by means of sight rails and boning rods.
2. Sketch and describe the class A, B, C and D methods of bedding of drainage pipelines and state the advantages of each.

3. State the advantages of flexible joints on drains over rigid joints.
4. Describe by the aid of sketches the following tests on drainage pipelines:
   (a) water test,
   (b) smoke test,
   (c) air test,
   (d) test for straightness and obstruction.
5. Explain the Building Regulations 1985 requirements for drain trenches adjacent to building foundations.
6. Sketch a longitudinal section through a petrol interceptor and explain its operation.
7. Sketch a plan of a drainage system for a large public vehicle garage.
8. Sketch a section through a grease trap suitable for drainage from a canteen kitchen and explain how the trap prevents grease entering the drainage system.
9. Sketch a section through a drainage pumping station when the motor room is to be
   (a) above ground level,
   (b) below ground level.
10. Sketch and describe the operation of a sump pump installation.

Chapter 9
Gradients, Rainwater Collection and Disposal

Gradients

For peak flows of more than 1 l/s but less than 2.5 l/s, a gradient not flatter than 1 in 70 is generally satisfactory if the drain serves the equivalent discharge of at least one WC. For peak flows of 2.5 l/s and more, gradients not flatter than 1 in 130 for a 100 mm diameter, or 1 in 200 for a 150 mm diameter drain may be used providing there is a high standard of workmanship.

For general purposes, gradients should not be flatter than 1 in 80 for a 100 mm diameter, or 1 in 150 for a 150 mm diameter drain. Where flows are small, or where continuous flows containing solid matter are less than 1 l/s, or when the drain is long, steeper gradients may be necessary, which generally should not be less than 1 in 40. Steep gradients increase the amount of excavation necessary, so for reason of economy flatter gradients are preferred.

Calculation of Velocity and Gradient

Various formulae and tables may be used to find the velocity of flow and the gradient for a drain. The best known formula, which may be used for both pipes and channels, is known as Chezy's formula, and is expressed as

$$v = c \sqrt{(mi)}$$

where v is the velocity of flow in m/s, c is the Chezy constant, m is the hydraulic mean depth in m, and i is the inclination or fall.

The Chezy constant may be found from the following formula:

$$c = \sqrt{\frac{2g}{f}}$$

where g is the acceleration due to gravity (9.81) and f is the coefficient of friction. The average coefficient of friction may be taken as 0.0064 and therefore the Chezy constant, C, would be;

$$c = \sqrt{\frac{2 \times 9.81}{0.0064}}$$
$$= 55$$

For pipes flowing half or full bore, the hydraulic mean depth is equal to $d/4$, which can be shown as follows. (See also Fig. 9.1.)

1. Half full bore
$$m = \frac{\text{Wetted area}}{\text{Wetted perimeter}}$$
$$= \frac{\pi r^2/2}{2\pi r/2}$$

by cancellation

$$m = \frac{r}{2}$$
$$= \frac{d}{4}$$

2. Full bore
$$m = \frac{\pi r^2}{2\pi r}$$

by cancellation

$$m = \frac{r}{2}$$

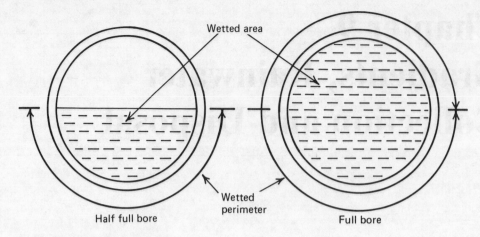

Fig. 9.1 Hydraulic mean depth for pipes.

Table 9.1 Values of m from the Diameter of Pipe

Depth of flow	Value of m
Full or half-full bore	Diameter × 0.25
$\frac{3}{4}$-depth of flow	Diameter × 0.30
$\frac{2}{3}$-depth of flow	Diameter × 0.29
$\frac{1}{3}$-depth of flow	Diameter × 0.19
$\frac{1}{4}$-depth of flow	Diameter × 0.15

$$= \frac{d}{4}$$

Table 9.1 gives the method of finding the value of m from the diameter of pipe.

Example 9.1. Calculate the velocity of flow through a 100 mm diameter drain flowing half full when the gradient is 1 in 60. (Chezy constant = 55.)

$$v = c \sqrt{\left(\frac{d}{4} \times \frac{1}{l} \right)}$$

$$= 55 \sqrt{\frac{0.1}{4} \times \frac{1}{60}}$$

$$= 55 \sqrt{(0.025 \times 0.0167)}$$

$$= 55 \times 0.02$$

$$= 1.1 \text{ m/s}$$

If it is required to find the discharge through the drain in l/s, this may be calculated as follows.

$$Q = v a$$

where Q is the volume of flow in m³/s, v is the velocity of flow in m/s, and a is the cross-sectional area of pipe in m².

$$Q = \frac{v \pi r^2}{2}$$

$$= \frac{1.1 \times 3.142 \times 0.05 \times 0.05}{2}$$

$$= 0.00432 \text{ m}^3/\text{s}$$

$$= 4.32 \text{ l/s}$$

It is often necessary to find the gradient.

Example 9.2. Calculate the gradient required for a 150 mm diameter drain flowing full bore when the velocity of flow is to be 1.5 m/s. (Chezy constant = 55.)

$$v = c \sqrt{(m\ i)}$$

$$= c \sqrt{\frac{d}{4} \times \frac{1}{l}}$$

By transposition

$$\left(\frac{v}{c} \right)^2 = \frac{d}{4} \times \frac{1}{l}$$

$$\left(\frac{v}{c} \right)^2 \times \frac{4}{d} = \frac{1}{l}$$

$$l = \left(\frac{c}{v} \right)^2 \times \frac{d}{4}$$

$$= \left(\frac{55}{1.5} \right)^2 \times \frac{0.150}{4}$$

$$= 1344.4443 \times 0.0375$$

$$= 50.42$$

Gradient = 1 in 50, approx.

146

It may sometimes be necessary to find the velocity of flow in a rectangular or square channel.

Example 9.3 Calculate the velocity of flow in the rectangular channel shown in Fig. 9.2 when laid to a gradient of 1 in 80. (Chezy constant = 55.)

$$m = \frac{b \times d}{b + 2d}$$
$$= \frac{0.5 \times 0.3}{0.5 + (2 \times 0.3)}$$
$$= \frac{0.15}{1.1}$$

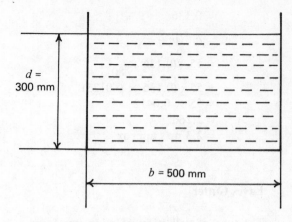

Fig. 9.2 Hydraulic mean depth for a channel.

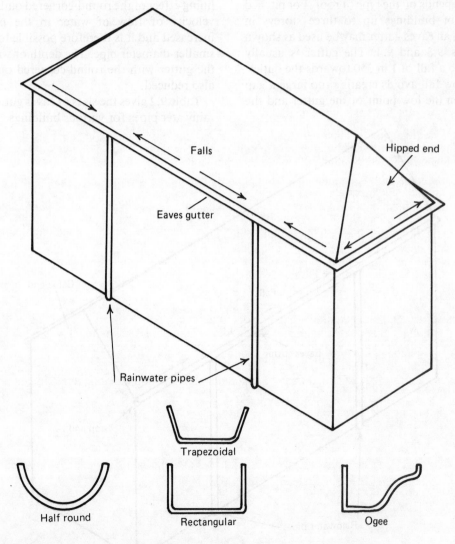

Falls

Hipped end

Eaves gutter

Rainwater pipes

Trapezoidal

Half round

Rectangular

Ogee

Eaves gutter sections

Fig. 9.3 Rainwater collection from a hipped-end roof.

$$= 0.136$$
$$v = c \sqrt{(m\,i)}$$
$$= 55 \sqrt{\left(0.136 \times \frac{1}{80}\right.}$$
$$= 55 \sqrt{(0.136 \times 0.0125)}$$
$$= 55 \times 0.0412$$
$$= 2.266$$
$$= 2.3 \text{ m/s approx.}$$

Eaves Gutter

The type of collection of rainwater from a roof depends on the type of roof. For pitched roofs of buildings up to three storeys in height, an eaves gutter may be used as shown in Figs. 9.3 and 9.4. The gutter is usually fixed to a fall of 1 in 350 towards the outlet. This low fall avoids creating too large a gap between the low point of the gutter and the edge of the roof, but it is sufficient to allow the flow of water and any slight settlement of the gutter.

Various gutter sections are obtainable, including half round, rectangular and ogee. The rectangular section permits a larger flow of water for the same width of gutter. The gutters are made from cast iron (which requires protection from corrosion), enamelled steel, aluminium alloy, uPVC and asbestos cement. Gutter outlets may be sharp cornered or round cornered. Figs. 9.5a and b show the discharge of water from sharp and round-cornered outlets. Due to the streamlining effect of the round-cornered outlet, the velocity of flow of water in the pipe is increased and it is therefore possible to use a smaller-diameter pipe. The depth of water in the gutter with the round-cornered outlet is also reduced.

Table 9.2 gives the sizes of eaves gutter and rainwater pipes for various buildings.

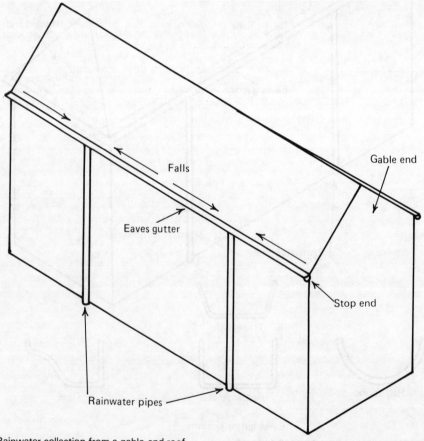

Fig. 9.4 Rainwater collection from a gable-end roof.

148

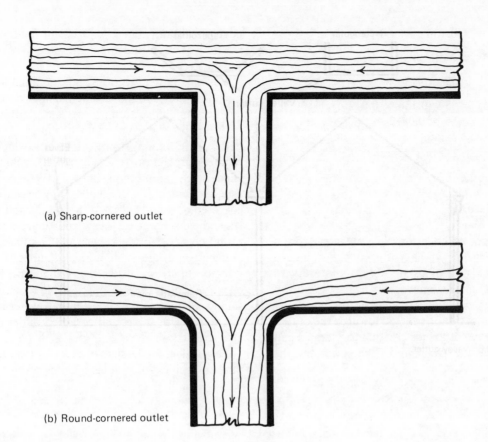

(a) Sharp-cornered outlet

(b) Round-cornered outlet

Fig. 9.5   Outlets from eaves gutter.

Table 9.2   Sizes of Eaves Gutter and Rainwater Pipes

Note: the minimum fall of the gutter is 1 in 600.

Diameter of gutter (mm)	Diameter of rain-water pipe (mm)	Application	Diameter of gutter (mm)	Diameter of rain-water pipe (mm)	Application
75	50	Domestic garages, garden sheds, greenhouses, dormer, bay windows	125	75	Large houses, offices, flats and shops, farm buildings, industrial buildings
100	63	Houses, flats, small shops and offices, garage blocks, site huts	150	100	Large roof areas of agricultural, commercial and industrial buildings, warehouses, super markets and stores

149

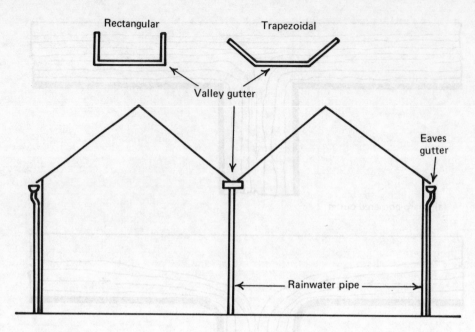

Fig. 9.6 Valley gutter.

Valley Gutter (Fig. 9.6)

This type of gutter is formed between two pitched roofs and will usually have to carry more water than an eaves gutter. The section may be rectangular or trapezoidal and it is essential to provide a minimum of 300 mm between the edges of the roofs to permit walking space along the gutter.

Parapet Wall Gutter (Fig. 9.7)

This is used for the pitched roofs buildings above three storeys in height and permits easier maintenance than an eaves gutter. The outlet from the gutter may be a chute, as shown in Fig. 9.8, or a catch pit as shown in Fig. 9.9. If required, a 40 mm bore storm overflow pipe may be fitted to the catch pit. The falls of both valley and parapet wall gutters are usually 1 in 80.

Flat Roofs

Various methods of rainwater collection may be used for flat roofs. Fig. 9.10a shows a flat roof drained by means of parapet gutters and Fig. 9.10b shows a flat roof drained without the use of gutters, which saves the construction of gutters but requires more falls on the roof towards the outlets. Fig. 9.10c shows a flat roof for a low-rise building where an eaves gutter may be used. The fall of the roof towards the gutter or outlets should not be less than 1 in 80 and preferably 1 in 60 to prevent the risk of 'ponding'.

Fig. 9.11 shows a bell-mouthed outlet for a flat roof. This type of outlet provides a streamlined effect to the flow of water: up to 140 m² of roof area may be drained by a 75 mm bore pipe, and up to 200 m² of roof area through a 100 mm bore pipe using this type of outlet.

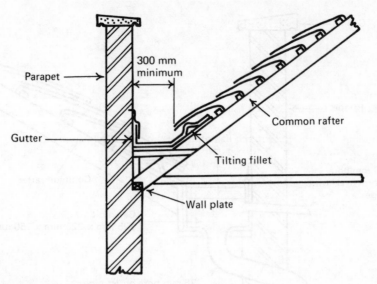

Fig. 9.7 Parapet wall gutter.

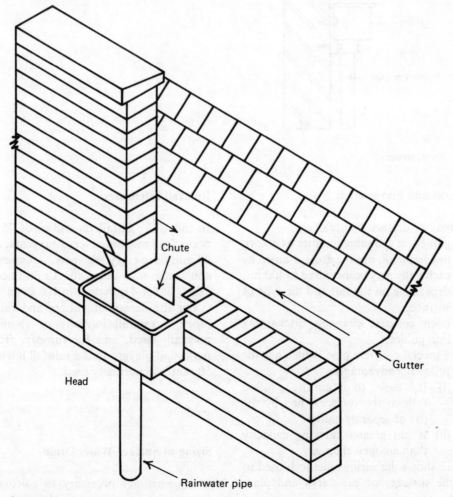

Fig. 9.8 Chute outlet.

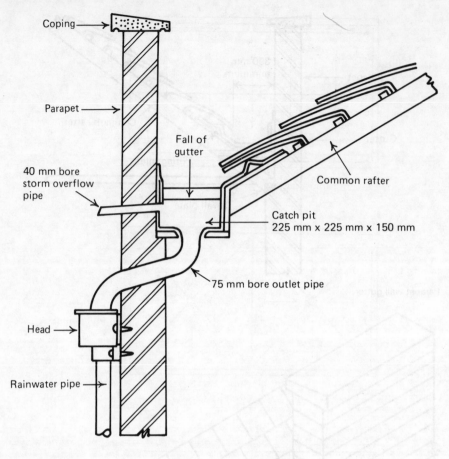

Coping

Parapet

Fall of gutter

40 mm bore storm overflow pipe

Common rafter

Catch pit 225 mm x 225 mm x 150 mm

75 mm bore outlet pipe

Head

Rainwater pipe

Fig. 9.9   Catch pit outlet.

Car Parks and Playgrounds

These may be drained by use of

1. gullies, in the same manner as shown in Chapter 8, but the areas drained by each gully may be increased to 400 m², depending on the fall and the type of surface,

2. open concrete channels discharging into gullies,

3. a special concrete pipe, which has the following advantages.
    (i) It is easier to obtain the required falls of the surfaces than by the use of separate gullies.
    (ii) It has greater carrying capacity than an open channel.

Fig. 9.12 shows the various methods used to drain the surfaces of car parks and playgrounds.

Rainfall Intensities

In the UK, rainfall intensities of 75 mm/h occur for 5 min once every four years, and for 20 min once every ten years. An intensity of 150 mm/h may occur for 3 min once in 50 years, or for 4 min once in 100 years.

For the design of gutters and rainwater pipes, a rainfall intensity of 75 mm/h is normally used, and for run-offs from car parks and playgrounds a rainfall intensity of 50 mm/h is normally used.

Sizing of Surface Water Drains

It is sometimes necessary to calculate the diameter of a surface water drain.

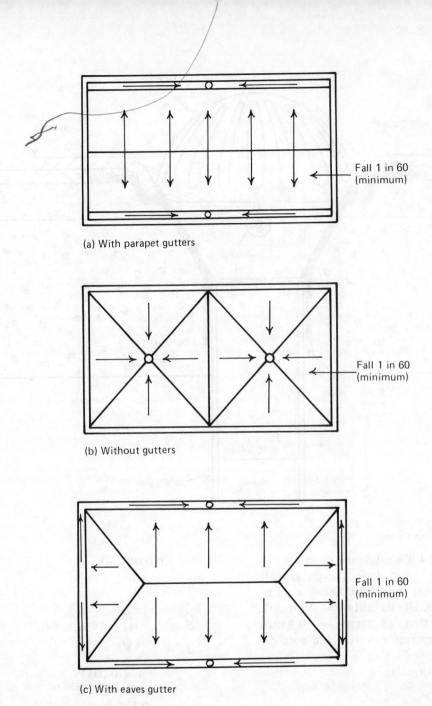

(a) With parapet gutters

Fall 1 in 60
(minimum)

(b) Without gutters

Fall 1 in 60
(minimum)

(c) With eaves gutter

Fig. 9.10 Flat roof.

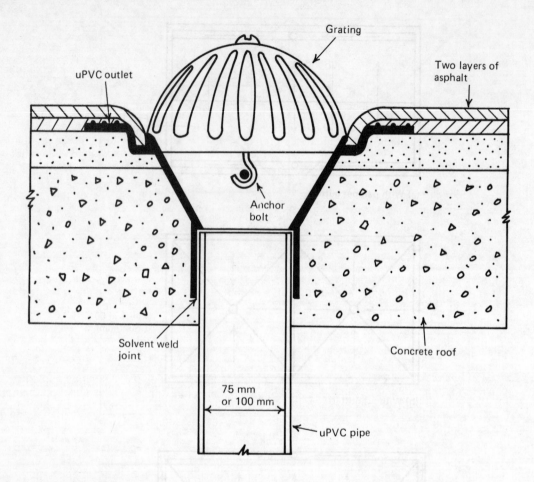

Fig. 9.11 Outlet to flat roof.

Example 9.4. Calculate the diameter of a main surface water drain for an asphalt-covered car park measuring 100 m × 75 m. (Assume rainfall intensity = 50 mm/h, velocity of flow of water = 0.8 m/s, impermeability factor = 0.9, full-bore discharge.)

Using the formula:

$$Q = va$$

where Q is the volume of flow in m³/s, v is the velocity of flow in m/s, and a is the cross-sectional area of pipe in m²,

$$Q = \frac{\begin{array}{c}\text{Area to}\\ \text{be}\\ \text{drained}\\ \text{(m}^2\text{)}\end{array} \times \begin{array}{c}\text{Rainfall}\\ \text{intensity}\\ \\ \text{(m/h)}\end{array} \times \begin{array}{c}\text{Imper-}\\ \text{meability}\\ \\ \text{factor}\end{array}}{3600}$$

$$= \frac{100 \times 75 \times 0.05 \times 0.9}{3600}$$

$$= 0.09375 \text{ m}^3/\text{s}$$

$$a = \frac{\pi d^2}{4}$$

$$Q = \frac{v \pi d^2}{4}$$

$$d = \sqrt{\frac{4 Q}{v \pi}}$$

$$= \sqrt{\frac{4 \times 0.09375}{0.8 \times 3.142}}$$

$$= 0.386 \text{ m}$$

$$= 386 \text{ mm}$$

Nearest pipe diameter = 400 mm.

Exercises

1. Calculate the velocity of flow of water in m/s through a 150 mm bore drain flowing full bore when laid to a gradient of 1 in 180. (Chezy constant = 55.)

154

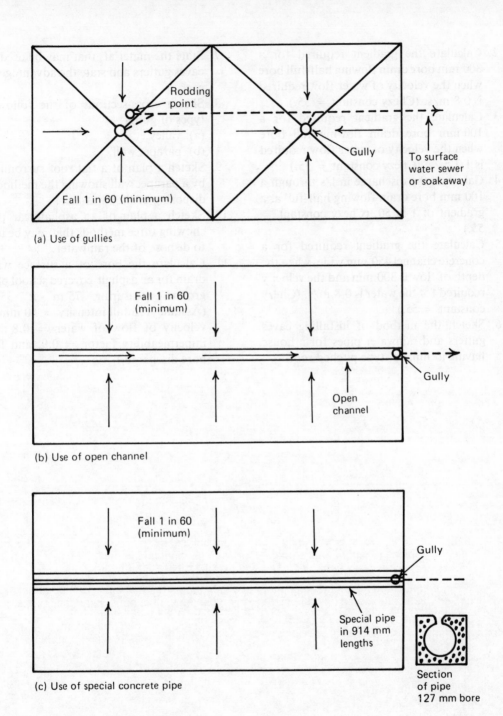

(a) Use of gullies

Fall 1 in 60 (minimum)

Rodding point

Gully

To surface water sewer or soakaway

Fall 1 in 60 (minimum)

Gully

Open channel

(b) Use of open channel

Fall 1 in 60 (minimum)

Gully

Special pipe in 914 mm lengths

(c) Use of special concrete pipe

Section of pipe 127 mm bore

Fig. 9.12   Drainage of car parks and playgrounds.

2. Calculate the gradient required for a 300 mm bore drain flowing half-full bore when the velocity of water flow required is 0.8 m/s. (Chezy constant = 55.)
3. Calculate the gradient required for a 100 mm bore drain flowing full bore when the velocity of water flow required is 1.2 m/s. (Chezy constant = 55.)
4. Calculate the discharge in l/s through a 100 mm bore drain flowing half-full at a gradient of 1 in 50. (Chezy constant = 55.)
5. Calculate the gradient required for a concrete channel 450 mm wide, when the depth of flow is 300 mm and the velocity required for the water is 0.8 m/s. (Chezy constant = 55.)
6. Sketch the method of installing eaves gutters and rainwater pipes for a house having a 'hipped'-type pitched roof.
7. State the materials that may be used for eaves gutters and state the advantages of each.
8. Sketch cross-sections of the following types of gutter:
   (a) valley,
   (b) parapet wall.
9. Sketch a plan of a flat roof surrounded by a parapet wall showing the method of disposing of the rainwater.
10. Sketch a plan of an asphalt car park showing three methods that may be used to dispose of the rainwater.
11. Calculate the bore of a surface water drain for an asphalt-covered school playground measuring 75 m × 35 m. (Assume rainfall intensity = 50 mm/h, velocity of flow of water = 0.8 m/s, impermeability factor = 0.9, and full-bore discharge.)

Answers to Numerical Exercises

Chapter 9

1. 0.794 m/s
2. 1 in 355
3. 1 in 53
4. 4.8 l/s
5. 1 in 608
11. 228 mm (Nearest pipe size = 250 mm bore)

Index